Deterrence in the Age of Thinking Machines

YUNA HUH WONG, JOHN M. YURCHAK, ROBERT W. BUTTON,
AARON FRANK, BURGESS LAIRD, OSONDE A. OSOBA,
RANDALL STEEB, BENJAMIN N. HARRIS, SEBASTIAN JOON BAE

For more information on this publication, visit www.rand.org/t/RR2797

Library of Congress Cataloging-in-Publication Data is available for this publication.
ISBN: 978-1-9774-0406-0

Published by the RAND Corporation, Santa Monica, Calif.
© Copyright 2020 RAND Corporation
RAND® is a registered trademark.

Cover: Getty Images Plus.

Support RAND
Make a tax-deductible charitable contribution at
www.rand.org/giving/contribute

www.rand.org

Preface

We expect artificial intelligence (AI) and autonomous systems to significantly change the future battlefield. Militaries around the world are incorporating AI and autonomy into their organizational processes, command and control systems, logistics systems, and, of course, weapon systems themselves—with an aim toward leveraging current developments from the commercial world. As AI and autonomy proliferate on the battlefield, an important question arises: How might deterrence be affected by the proliferation of AI and autonomous systems? Up to this point, deterrence has primarily involved humans attempting to affect the decision calculus and perceptions of other humans. But what happens when decisionmaking processes are no longer fully under the control of humans? In this report, we lay out some initial considerations and present ideas for how deterrence could change in the age of AI and autonomy.

RAND Ventures

The RAND Corporation is a research organization that develops solutions to public policy challenges to help make communities throughout the world safer and more secure, healthier and more prosperous. RAND is nonprofit, nonpartisan, and committed to the public interest.

RAND Ventures is a vehicle for investing in policy solutions. Philanthropic contributions support our ability to take the long view, tackle tough and often-controversial topics, and share our findings in innovative and compelling ways. RAND's research findings and recommendations are based on data and evidence, and therefore do not necessarily reflect the policy preferences or interests of its clients, donors, or supporters.

Funding for this venture was made possible by the independent research and development provisions of RAND's contracts for the operation of its U.S. Department of Defense federally funded research and development centers.

International Security and Defense Policy Center

This research was conducted within the International Security and Defense Policy Center of the RAND National Defense Research Institute, a federally funded research and development center sponsored by the Office of the Secretary of Defense, the Joint Staff, the Unified Combatant Commands, the defense agencies, and the defense Intelligence Community.

For more information on the International Security and Defense Policy Center, see www.rand.org/nsrd/ndri/centers/isdp or contact the director (contact information is provided on the webpage).

Contents

Preface iii
Figures and Tables vii
Summary ix
Acknowledgments xv
Abbreviations xvii

CHAPTER ONE
Introduction 1
Background 2
Why Is This Topic Important? 5
Approach 7
How to Read This Report 8

CHAPTER TWO
Key Deterrence Concepts 11
Deterrence 11
Escalation 13
Stability and Instability 15
Security Dilemma 15

CHAPTER THREE
Artificial Intelligence and Autonomous Systems 17
Artificial Intelligence 17
Autonomy 22

CHAPTER FOUR
Potential Futures in a World of Proliferated Artificial Intelligence
    and Autonomous Systems 27
Key Factors Affecting Deterrence in the Age of Thinking Machines 27
Creating One Future World 36

CHAPTER FIVE
A Wargame of Artificial Intelligence and Autonomous Systems 39
Wargame Overview... 39
Wargame Events... 43
Wargame Limitations... 48

CHAPTER SIX
Wargame Insights and Debates 51
Wargame Insights... 51
Debates During the Game 53

CHAPTER SEVEN
Implications for Deterrence.. 59
Deterrence Concepts Revisited... 59
Comparisons with Nuclear Weapon Proliferation 61
How Escalatory Dynamics May Change 63

CHAPTER EIGHT
Implications for Decisionmaking 69
Inadvertent Engagement 69
Trust and Autonomy.. 74
Time Scale and Decisionmaking... 75
Considerations When Building Autonomous Forces 80

CHAPTER NINE
Conclusion and Areas for Further Research............................. 83
Conclusion.. 83
Areas for Further Research... 83

APPENDIX
General Morphological Analysis... 87

References 93

Figures and Tables

Figures

7.1. Understanding Required in Traditional Deterrence............67
7.2. Understanding Required in Deterrence with AI and
Autonomous Systems...67
8.1. Boyd's Observe, Orient, Decide, Act (OODA) Loop...........76
8.2. Reaction Speeds and Type of Autonomy.........................78

Tables

4.1. One Future World...37
5.1. Relative Distribution of Military Assets........................42
7.1. Nuclear Weapon Proliferation Versus Autonomous System
Proliferation...62
7.2. Human and Machine Configurations and Potential
Escalatory Dynamics...64
7.3. Potential Machine Misinterpretation of Human Signaling....66
8.1. Inadvertent Engagements by Autonomous Systems............72
8.2. Potential Advantages and Disadvantages of Military
Autonomous Systems..73
A.1. Morphological Field..89
A.2. Cross-Consistency Analysis.....................................90
A.3. One Future World...92

Summary

Artificial intelligence (AI) and autonomous systems have the potential to change the future of warfare. The increased use of unmanned systems on the battlefield, breakthroughs in commercial AI, and interest that many countries now have in AI and autonomous systems for military applications make it likely that such systems will be widely used in future conflicts. Yet what happens to deterrence and escalation when decisions can be made at machine speeds and when states can put fewer human lives at risk?

In this exploratory report, we discuss key deterrence concepts, offer a brief background on AI and autonomy, identify key factors that may shape deterrence and escalation as use of these systems increases, present a wargame in which several countries with AI and advanced autonomous systems confronted one another, offer potential implications that these technologies have for deterrence and escalation, and identify areas for further research.

A Wargame of Artificial Intelligence and Autonomy

Central to our examination of this topic was a wargame involving the United States, China, Japan, South Korea, and North Korea in a future world with AI and autonomous forces. In the wargame scenario, China was the global power, and the United States, Japan, and South Korea remained allies countering China.

The purpose of the wargame was to have a structured conversation in an operational framework about how AI and autonomy could affect deterrence as events unfolded. The goal of the game was not to test or discover how players could prevail against others using these technologies. Instead, the goal was to explore the ways they could affect escalation and deterrence.

Wargame Summary

The wargame began with China attempting to exert greater control in the region and the United States and Japan resisting this attempt. The game escalated at several points, first into conflict between unmanned systems, then eventually into one in which Chinese and U.S. military personnel were killed. There was both intentional and inadvertent escalation. The United States and Japan engaged in joint exercises and deliberately sought to provoke the Chinese. China took the escalatory step of declaring unrestricted submarine warfare in order to enforce a blockade of certain Japanese ports, and it sank an unmanned Japanese cargo ship. U.S. and Japanese antisubmarine warfare assets then sank a manned Chinese submarine, which represented the first human casualties in the game. The U.S. and Japan players were unable to deescalate the situation at this point. China retaliated with a missile attack against the U.S. and Japanese fleet, also causing human casualties. The game ended with the crisis still escalating.

Wargame Insights

Although this was only a single wargame, there were several interesting, initial insights.

Manned systems may be better for deterrence than unmanned ones. While the U.S. and Japanese systems were unmanned in this scenario, the Chinese had some manned platforms. The presence of humans on Chinese platforms made the U.S. and Japan players more hesitant to use force and often put the onus on them to look for offramps to avoid further escalation.

Replacing manned systems with unmanned ones might not be seen as a reduced security commitment. Manned systems may be better than unmanned ones for deterrence, but replacing manned U.S. systems with unmanned ones was not always seen as a sign of reduced commitment by U.S. allies in the region.

Players put their systems on different autonomous settings to signal resolve and commitment during the conflict. Deliberately taking actions and decisions out of human hands and going to "full auto" emerged as a way to show that players were willing to use force.

The speed of autonomous systems did lead to inadvertent escalation in the wargame. Setting forces on "full auto" to signal resolve did in one case lead to inadvertent escalation. Systems set to autonomous mode reacted with force to an unanticipated situation in which the humans did not intend to use force.

Other insights from the game were that distances still mattered, that the presence of AI in decisionmaking created opportunities to confuse players and interject uncertainty into dynamics, that countries in the game had a mix of different human-and-machine architectures at different levels, and that players can dramatically overestimate adversary modernization in unobservable areas such as AI.

Implications for Deterrence

Autonomous and unmanned systems could affect extended deterrence and our ability to assure our allies of U.S. commitment. Allies in this specific wargame did not feel that replacing U.S. troops with robots was a sign of reduced commitment, but this may not always be the case. Autonomous systems could increase the credibility of U.S. conventional, extended deterrence if allies perceive these systems as more capable, or if they perceive U.S. leaders as more likely to employ autonomous systems because they reduce risk to U.S. personnel. Alternatively, allies might see U.S. reliance on unmanned systems as an unwillingness to put American lives on the line during confrontations with adversaries. This could reduce the United States' ability to assure its allies.

Widespread AI and autonomous systems could lead to inadvertent escalation and crisis instability. Decisions made at machine rather than human speeds also have the potential to escalate crises at machine speeds. During protracted crises and conflicts, there could be strong incentives for each side to use autonomous capabilities early and extensively to gain military advantage. This raises the possibility of first-strike instability. AI and autonomous systems also have the potential to reduce strategic stability between adversaries and make the use of force more likely as they lower the risks to military personnel. An arms race in autonomous systems between the United States and China already appears imminent and is likely to increase instability. A proliferation of autonomous systems could also ignite a serious search for countermeasures and exacerbate uncertainties that leave countries perceiving themselves as less secure, in a textbook example of the security dilemma.

Different mixes of human and artificial agents could affect the escalatory dynamics between two sides. In our wargame, U.S. and Japanese forces went down a path of using largely unmanned systems with human control, whereas Chinese forces included manned systems controlled by machines. This created a dynamic in which U.S. and Japanese forces tried several times to deescalate the crisis because of their reluctance to kill Chinese personnel, whereas the player for the Chinese forces did not appear to feel the need to deescalate.

From this, we hypothesize a larger dynamic whereby different configurations of humans and machines—in terms of both physical presence and decisionmaking—may affect escalatory dynamics. First, we hypothesize that when physical presence and decisionmaking are both primarily human, there is a lower escalatory dynamic (due to slower, human decisionmaking) but a higher cost to miscalculation (because human lives are at stake). Second, we postulate that when the physical presence is primarily machine and decisionmaking is primarily human, as U.S. forces were in our wargame, there is both a lower escalatory dynamic and a lower cost of miscalculation (because human lives are not at stake). For the third case, where physical presence is primarily human but the decisionmaking is primarily machine, as were Chinese forces in the wargame, we hypothesize a higher escalatory dynamic (due to machine decisionmaking) and a higher cost of miscalculation. Lastly, we hypothesize that when both the physical presence and the decisionmaking are primarily machine, there is a higher escalatory dynamic but a lower cost of miscalculation (because human lives are not at stake). *When adversaries have different configurations, the onus may be more on one to deescalate.*

Machines will likely be worse at understanding the human signaling involved in deterrence, especially deescalation. AI that is programmed to aggressively exploit tactical and operational advantages could misunderstand adversary attempts to signal resolve but avoid signaling imminent attack, to avoid further escalation, to or actively deescalate a situation.

Whereas traditional deterrence has largely been about humans attempting to understand other humans, deterrence in this new age involves understanding along a number of additional pathways. Not only must humans understand adversary humans, they must adequately understand their own machines and adversary machines. Machines must also accurately understand their own humans, adversary humans, and adversary machines. These additional pathways introduce possibilities for misinterpretation, misperception, and miscalculation. Continual modernization over time means that countries will likely always have a mix of new and legacy systems, potentially making it difficult to understand the full set of human-machine interactions within any military.

Past cases of inadvertent engagement of friendly or civilian targets by autonomous systems may offer insights about the technical accidents or failures involving more-advanced systems. Military autonomous systems are not new, and neither is inadvertent engagement by such autonomous systems as landmines, torpedoes, the Phalanx antimissile system, and the Aegis weapon system. Human or system target misidentification was one common problem in many of these historical cases. It is possible that AI could improve target identification and reduce this type of error. It is also possible that human error interacting with even more-complex systems could contribute to future inadvertent engagement.

Areas for Future Research

Conduct further work on deterrence theory and other frameworks to explicitly consider the potential effects of AI and autonomous systems. This report presents an initial look at the topic. Further thought experiments and analyses are necessary before we will be able to successfully manage intended deterrence and avoid unintended escalation.

Evaluate the escalatory potential of new systems. We recommend system-level review of proposed AI implementations to assess any escalatory implications that should be considered in its design, development, testing, or use.

Evaluate the escalatory potential of new operating concepts. How might decision cycles, processes, and operating concepts that use AI and autonomous systems contribute to miscalculation or inadvertent escalation? Does a given concept allow for deescalation? How? Are certain operating concepts inherently more escalatory even as they offer operational advantages?

Wargame additional scenarios at the operational and strategic levels. We gained insights from a single wargame; additional wargames across a variety of scenarios, with different adversaries and allies, may offer additional insights. Strategic-level wargaming over greater time horizons may yield insights on topics such as horizontal escalation or arms-race instability.

Acknowledgments

We are especially indebted to Susan L. Marquis, dean of the Pardee RAND Graduate School; Jack Riley, vice president and director of the RAND National Security Research Division; and Howard Shatz, former director of RAND-Initiated Research, for their interest in and support for our work on this topic. We also thank International Security and Defense Policy Center director Christine Wormuth, former acting director Andrew Parasiliti, associate director Michael McNerney, and acting associate director Richard Girven for their oversight of this project.

We thank T.J. Gilmore, a Navy fellow at RAND, for participating in our workshop and wargame and providing valuable insights.

We are also grateful to Paul Scharre at the Center for a New American Security and Edward Geist at RAND for their thoughtful and constructive reviews of a previous draft of this report. Finally, we thank RAND communications analyst Barbara Bicksler, who helped us to better organize our report and to make several of our points clearer.

Abbreviations

A2/AD	anti-access/area denial
AI	artificial intelligence
ALFUS	Autonomy Levels for Unmanned Systems
ASW	antisubmarine warfare
DARPA	Defense Advanced Research Projects Agency
DoD	U.S. Department of Defense
GMA	general morphological analysis
HITL	humans in the loop
HOOTL	humans out of the loop
HOTL	humans on the loop
ISR	intelligence, surveillance, and reconnaissance
NIST	National Institute of Standards and Technology
OODA	observe, orient, decide, act
UAV	unmanned aerial vehicle

CHAPTER ONE

Introduction

Artificial intelligence (AI) and autonomous systems have the potential to significantly change the future battlefield. Within this broad topic, the specific focus of this exploratory examination is how AI and autonomy may affect deterrence and escalation in crises and conflicts. In this report, we lay out considerations and present ideas for how deterrence could change in the age of AI and autonomy.

Our work addressed a number of key research questions:

- What are the implications of adding thinking machines and autonomous systems to the practices that countries have developed to signal one another about the use of force and its potential consequences?
- What happens to deterrence and escalation when decisions can be made at machine speeds and are carried out by forces that do not risk human lives of the using state or actor?
- How might the rise of these capabilities weaken or strengthen deterrence?
- What are potential areas of miscalculation and unintended consequences?

These questions arise as the United States and other states around the world actively seek to develop and incorporate AI and autonomy into their military forces. The visible advances that AI and autonomy are making in the commercial sector have prompted interest and excitement about bringing similar changes and advances to the defense and intelligence sectors. In this report, we do not focus on whether these technologies should be used in military applications, or how best to implement them for national security purposes. We take it as a given that AI and autonomy will be increasingly adopted by the world's militaries and deployed in ways that maximize their military benefits. We instead focus on the implications of the wider adoption of these technologies on deterrence.

Background

Three trends bring us to where we are today—a world where militaries are seeking to use AI and autonomous systems. The first trend became notable after September 11, 2001, and the start of conflicts in Iraq and Afghanistan: the rise of unmanned systems on the battlefield. These conflicts prompted proliferation of unmanned aerial systems, which have included unarmed intelligence, surveillance, and reconnaissance (ISR) drones as well as armed drones used for precision strike missions.[1] The introduction and use of thousands of ground robots, particularly to counter improvised explosive devices, also arose during this period.[2] These unmanned systems were mainly operated remotely by humans and incorporated little autonomous decisionmaking, but they set the stage for additional thinking about robots in war.

The second major trend has been advances in AI in areas such as computer vision, AI planning, machine learning, natural language processing, and robotics.[3] Autonomous vehicle technology has also been a highly visible area, with the Defense Advanced Research Projects

---

[1] Matthew Fuhrmann and Michael C. Horowitz, "Droning On: Explaining the Proliferation of Unmanned Aerial Vehicles," *International Organization*, Vol. 71, No. 2, Spring 2017, pp. 397–418; Michael J. Boyle, "The Race for Drones," *Orbis*, November 24, 2014, pp. 76–94; Lynn E. Davis, Michael J. McNerney, James Chow, Thomas Hamilton, Sarah Harting, and Daniel Byman, *Armed and Dangerous? UAVs and U.S. Security*, Santa Monica, Calif.: RAND Corporation, RR-449-RC, 2014.

[2] P. W. Singer, *Wired for War: The Robotics Revolution and Conflict in the 21st Century*, New York: Penguin Books, 2009, pp. 19–23.

[3] Stanford University, One Hundred Year Study on Artificial Intelligence, *Artificial Intelligence and Life in 2030: Report of the 2015 Study Panel*, September 2016.

Agency (DARPA) holding three "Grand Challenges" between 2003 and 2007 that accelerated autonomous vehicle technology development.[4] Google's self-driving car initiative prompted similar investment in such technology by Audi, Toyota, Ford, Tesla, BMW, and Uber.[5] Using machine learning, Google DeepMind's AlphaGo defeated 18-time world go champion Lee Sedol in a 2016 match in South Korea. AlphaGo then defeated world champion Ke Jie the following year in China.[6] Worldwide, companies invested an estimated $26–39 billion in AI in 2016 across a number of sectors in the economy.[7]

The third trend is that major powers are seeking to accelerate the use of AI autonomous systems in their militaries. Former U.S. Secretary of Defense James Mattis expressed interest in AI throughout his tenure, even surmising whether AI might change the fundamental nature of war.[8] Unclassified reports of U.S. defense spending indicate an increase in investments in AI, with the U.S. Department of Defense (DoD) spending $7.4 billion in 2017 on AI and supporting areas, up from expenditures of $5.6 billion in 2012.[9] The U.S. also created a Joint Artificial Intelligence Center in 2018 with oversight of

---

[4] James M. Anderson, Nidhi Kalra, Karlyn D. Stanley, Paul Sorensen, Constantine Samaras, and Oluwatobi A. Oluwatola, *Autonomous Vehicle Technology: A Guide for Policymakers*, Santa Monica, Calif.: RAND Corporation, RR-433-2-RC, 2016, pp. 56–57.

[5] Anderson et al., p. 57; Charlotte Jee and Christina Mercer, "Driverless Car News: The Great Driverless Car Race: Where Will the UK Place?" *Tech World*, November 20, 2017.

[6] Christof Koch, "How the Computer Beat the Go Master," *Scientific American,* March 19, 2016; Paul Mozur, "Google's AlphaGo Defeats Chinese Go Master in Win for A.I.," *New York Times*, May 23, 2017; *AlphaGo*, directed by Greg Kohs, distributed by Moxie Pictures and Reel as Dirt, 2017.

[7] Jacques Bughin, Eric Hazan, Sree Ramaswamy, Michael Chui, Tera Allas, Peter Dahlstrom, Nicolaus Henke, and Monica Trench, *Artificial Intelligence: The Next Digital Frontier?* McKinsey Global Institute discussion paper, June 2017, p. 5.

[8] Aaron Mehta, "AI Makes Mattis Question 'Fundamental' Beliefs About War," *C4ISRNet*, February 17, 2018.

[9] Julian E. Barnes and Josh Chin, "The New Arms Race in AI," *Wall Street Journal*, March 2, 2018.

many defense AI efforts.[10] Spurred by both U.S. defense interest in AI and AlphaGo's victory in China, China announced its "Next Generational Artificial Intelligence Development Plan" in 2017, with the intention of becoming a dominant force in AI by 2030.[11] China has expressed fears of a "generational gap" between its military capabilities and U.S. capabilities, creating an incentive for China to closely track U.S. defense developments in AI.[12] Russia is also interested in developing AI for military use. The Russian Ministry of Defence announced a ten-point plan in 2018 outlining key public-private partnerships and next steps in research and development.[13] Russia is also expected to unveil its national AI strategy in 2019.[14]

This report thus proceeds on the assumption that the future battlefield will be one in which AI and autonomous military systems will be ubiquitous. As the third decade of the 21st century approaches, a growing number of states are engaged in well-funded efforts to develop and field AI and autonomous military systems.[15]

Some militaries are investing in such capabilities to improve a range of noncombat support and administrative functions, including cyber security, logistics, accounting, travel, and health care, and simply

---

[10] Sydney J. Freedberg, Jr., "Joint Artificial Intelligence Center Created Under DoD CIO," *Breaking Defense*, June 29, 2018b.

[11] Barnes and Chin, 2018; and Pablo Robles, "China Plans to Be a World Leader in Artificial Intelligence by 2030," *South China Morning Post*, October 1, 2018.

[12] Elsa B. Kania, *Battlefield Singularity: Artificial Intelligence, Military Revolution, and China's Future Military Power*, Washington, D.C.: Center for a New American Security, 2017, p. 5. For a contrary finding, see Lora Saalman, "Fear of False Negatives: AI and China's Nuclear Posture," *Bulletin of the Atomic Scientists*, April 24, 2018.

[13] Samuel Bendett, "Here's How the Russian Military Is Organizing to Develop AI," *Defense One*, July 20, 2018b.

[14] Samuel Bendett, "Putin Orders Up a National AI Strategy," *Defense One*, January 31, 2019.

[15] Barnes and Chin, 2018; Michael C. Horowitz, Gregory C. Allen, Edoardo Saravalle, Anthony Cho, Kara Frederick, and Paul Scharre, *Artificial Intelligence and International Security*, Washington, D.C.: Center for a New American Security, July 2018; and Tom Simonite, "For Superpowers, Artificial Intelligence Fuels New Global Arms Race," *WIRED*, September 8, 2017.

to stay abreast of beneficial applications made possible by advancements in AI and autonomy.[16] Other militaries, most notably those of China and Russia, are making major investments to keep up with and potentially offset U.S. autonomous military capabilities and U.S. attempts to obtain significant military advantages.[17]

Why Is This Topic Important?

Decades of deterrence research, significant amounts of it done at RAND,[18] rest on the fundamental premise that much of deterrence is about humans deterring other humans: humans sending and receiving signals with other humans, humans putting human lives at risk, and humans trying to understand what other humans may be thinking and anticipating their actions. But what happens with the addition of AI—that is, when some of the intelligence involved in signaling and deciding on both sides is no longer human? Related but separate, what happens to escalation and conflict management with the addition of unmanned systems, when fewer lives are perceived to be at risk?

This topic is important for several reasons. First, the likely development of more-advanced autonomous systems and the current proliferation of remotely operated unmanned systems indicate how rapid and widespread autonomous system proliferation might be. Advanced autonomous systems capable of performing increasingly complex missions without risking human operators will likely influence how states seek to deter opponents from undertaking aggression and how they coerce, compel, and attack their opponents. We expect the substantial presence of more-advanced autonomous systems on the battlefield to also reshape military doctrine and operational concepts.

---

[16] Itai Barsade and Michael C. Horowitz, "Artificial Intelligence Beyond the Superpowers," *Bulletin of the Atomic Scientists*, August 16, 2018.

[17] On China's efforts, see Kania, 2017; on Russia's efforts, see Samuel Bendett, "Russia Is Poised to Surprise the U.S. in Battlefield Robotics," *Defense One*, January 25, 2018a.

[18] Austin Long, ed., *Deterrence—From Cold War to Long War: Lessons from Six Decades of RAND Research*, Santa Monica, Calif.: RAND Corporation, MG-636-OSD/AF, 2008.

Second, the greatly increased decision-action speeds associated with AI and autonomous systems have the potential to escalate conflict much more quickly. An anticipated benefit of these systems is that they may be able to take calculated actions more quickly and consistently than a human decisionmaking cycle.[19] Such decision-action speed could give decisive operational advantage in contests against opponents who lack similar capabilities. Yet these same faster decision-action speeds could also escalate conflict much more rapidly than humans alone would, particularly if both sides are relying on machines to accelerate decision-action cycles.

Third, there may be unintended consequences of using more unmanned forces (whether they have autonomy or not), and of the perception that using unmanned forces instead of manned ones reduces risks to human life. Militaries may perceive that putting fewer of their personnel at risk through the use of unmanned systems is a good thing. However, does reducing the human cost of conflict to one or more sides make conflict or escalation more likely? And does putting fewer lives at risk through more unmanned forces mean the same to allies who are concerned about commitments?

Another reason this topic is important is that adding machine intelligence to the battlefield may also multiply the chances for misperception, miscalculation, and error. This concern grows as systems become increasingly complex.[20]

We already see the contours of different national philosophies for how the major militaries around the world may use these technologies during war. This is particularly true for the role of such technologies in human decisionmaking. For example, U.S. thinking and policy on the use of autonomous systems emphasizes the importance of keeping humans in the loop (HITL)—that is, involving a human

---

[19] Frans Osinga, *Science, Strategy and War: The Strategic Theory of John Boyd*, The Netherlands: Eburon Academic Publishers, 2005; and John Boyd, *A Discourse on Winning and Losing*, edited by Grant T. Hammond, Maxwell Air Force Base, Ala.: Air University Press, 2018.

[20] Sydney J. Freedberg Jr., "Why a 'Human in the Loop' Can't Control AI: Richard Danzig," *Breaking Defense*, June 1, 2018a.

in final decisions about the use of lethal force.[21] U.S. thinking is influenced by individuals such as former Deputy Secretary of Defense Bob Work, who articulated the concept of the "centaur," with humans and machines working together.[22]

On the other hand, some Chinese military thinkers have made reference to the concept of the "singularity": the point at which humans cannot keep up with the speed of decisionmaking and combat on the battlefield.[23] U.S. thinkers also perceive the Chinese as more willing to take humans out of the loop (HOOTL)—presumably to gain decision-action speed advantages in combat[24] because the Chinese potentially feel less constrained[25] and because of recent steps to centralize authority.[26] It is not yet clear whether this is actually the Chinese approach. However, we can and should begin thinking about the particular escalatory dynamics that may result when countries with different philosophies of use find themselves in conflict.

Approach

To perform our research, we assembled a diverse study team with expertise in areas such as deterrence, AI, U.S. military operations, force development, force posture, defense robotics, information science, social computation, wargaming, and problem-structuring methods. Other areas that we acknowledge are clearly relevant, but that we

---

[21] U.S. Army Training and Doctrine Command, Army Capabilities Integration Center, *The U.S. Army Robotic and Autonomous Systems Strategy*, Fort Eustis, Va., 2017, p. 3; and Department of Defense Directive 3000.09, *Autonomy in Weapons Systems*, Washington, D.C.: U.S. Department of Defense, November 21, 2012.

[22] Sydney J. Freedberg Jr., "Centaur Army: Bob Work, Robotics, and the Third Offset Strategy," *Breaking Defense*, November 9, 2015.

[23] Kania, 2017, p. 5.

[24] Kania, 2017, p. 5.

[25] Barnes and Chin, 2018.

[26] Joel Wuthnow and Phillip C. Saunders, *Chinese Military Reports in the Age of Xi Jinping: Drivers, Challenges, and Implications*, Washington, D.C.: Institute for National Strategic Studies, National Defense University, Chinese Strategic Perspectives 10, 2013, p. 33.

were unable to include within the scope of this project, include human-computer interaction and cognitive science. The primary focus of our work here was on conventional deterrence. Other RAND work has addressed the implications of AI for nuclear forces.[27]

The first step of our approach was to baseline the team's understanding of contemporary deterrence and AI theory and practice. Next, we conducted a weeklong workshop with the study team, capped with a wargame to identify and explore the major factors that we assess will likely affect deterrence. We created a potential future world, wargamed a scenario in Northeast Asia, and then spent the final day of our weeklong workshop identifying insights and ideas that had emerged during the course of our discussions.

Human Subjects Protection

We followed human subjects protection (HPS) protocols in the course of this study, in accordance with the appropriate statutes and U.S. Department of Defense (DoD) regulations governing HSP. The views of sources rendered anonymous in accordance with HSP protocols are solely their own and do not represent the official policy or position of DoD or the U.S. government.

How to Read This Report

The target audience of this report is a general reader who might not be an expert in either deterrence or AI but who is interested in some initial thoughts on the topic. The first part of the report attempts a concise overview of important background topics for those unfamiliar with them:

- Chapter Two provides working definitions of key deterrence concepts.
- Chapter Three provides a brief overview of AI and autonomy.

---

[27] Edward Geist and Andrew J. Lohn, *How Might Artificial Intelligence Affect the Risk of Nuclear War?* Santa Monica, Calif.: RAND Corporation, PE-296-RC, 2018.

The next two chapters provide a detailed exposition on how we approached thinking about the topic:

- Chapter Four contains the future factors we identified as potentially important and a description of a potential future that drew our interest.
- Chapter Five provides a description of the wargame scenario and the major events that transpired.

The remaining chapters lay out insights and thoughts about how AI and autonomous systems may impact escalation and deterrence:

- Chapter Six contains insights from the wargame and the debates that arose among members of the study team during the wargame.
- Chapter Seven discusses more general implications for deterrence in a broader context than our specific wargame.
- Chapter Eight examines additional implications for decisionmaking, deterrence, and escalation.
- Chapter Nine offers conclusions and areas for future research.

For readers who are more analytically oriented, we also include an appendix on the problem-structuring method we used to create the future world in Chapter Four.

CHAPTER TWO

Key Deterrence Concepts

What are we referring to when we discuss deterrence and escalation? What key concepts are important to know before we can consider implications of a future battlefield with widespread AI and autonomous systems? In this chapter, we present some working definitions of the most important concepts discussed in this report.

Deterrence

We define **deterrence** as a situation in which one state presents an opponent with a threat—implicit or explicit—designed to discourage the opponent from taking some proscribed aggressive action that it might otherwise consider taking but has not yet actually taken. Whether deterrence is **credible** depends upon three factors: (1) the adversary's perception of the **capability** of the deterrer to carry out the threatened punishment or denial of aims, (2) the adversary's perception of the **will** or **resolve** of the deterrer to make good on its threat, and (3) whether the deterrent threat is clearly **communicated** and understood by the adversary, a matter that previous crises have demonstrated cannot be taken for granted.

When the possibility of the opponent undertaking aggression is remote because no crisis is present,[1] or because no unusual military

---

[1] As Brecher and Wilkenfeld explain, a crisis can be understood as a situation in which three necessary and sufficient conditions exist. The highest-level decisionmakers of a state or two or more states must believe (1) that a threat to one or more basic values exists, (2) that there

preparations appear to be underway, then **general deterrence** is said to be at work. When a crisis is brewing or underway between two states, then **immediate deterrence** is said to be at work.[2]

During the Cold War, deterrence came to be widely understood almost solely in nuclear terms, or as **nuclear deterrence**. The United States most sought to deter nuclear aggression, and nuclear weapons were the ultimate means it threatened to employ to deter such aggression. However, the United States also sought to deter all major aggression by its adversaries and developed robust conventional military forces to underwrite its **conventional deterrence**.[3] Today, as during the Cold War, the United States has deterrence strategies in place that are designed to deter aggression against U.S. territory (**central deterrence**) and against U.S. allies in Europe and East Asia (**extended deterrence**). Deterrent threats regarding aggression against a power's homeland are "inherently credible," whereas threats to act in the event of aggression against a power's allies "have to be made credible."[4] Importantly, the threat has to be made credible not only to the adversary but also to the power's allies in order to **assure** allies.[5] As former British defense minister Denis Healey famously remarked, it takes "only five percent cred-

---

[2] Michael J. Mazarr, *Understanding Deterrence*, Santa Monica, Calif.: RAND Corporation, PE-295, 2018, p. 4.

is a finite time for response to the value threat, and (3) that the situation is characterized by a heightened probability of involvement in military hostilities. See Michael Brecher and Jonathan Wilkenfeld, *A Study of Crisis*, Ann Arbor, Mich.: University of Michigan Press, 1997, pp. 4–5.

[3] Michael S. Gerson, "Conventional Deterrence in the Second Nuclear Age," *Parameters*, Autumn 2009, pp. 32–48; John Stone, "Conventional Deterrence and the Challenge of Credibility," *Contemporary Security Policy*, Vol. 33, No. 1, 2012, pp. 108–123; Edward Rhodes, "Conventional Deterrence," *Comparative Strategy*, Vol. 19, No. 3, 2000, pp. 221–253; and Richard J. Harknett, "The Logic of Conventional Deterrence and the End of the Cold War," *Security Studies*, Vol. 4, No. 1, Autumn 1994, pp. 86–114.

[4] Thomas C. Schelling, *Arms and Influence*, New Haven, Conn.: Yale University Press, 1966, p. 36.

[5] Linton Brooks and Mira Rapp-Hooper, "Extended Deterrence, Assurance, and Reassurance in the Pacific During the Second Nuclear Age," in Ashley J. Tellis, Abraham M. Denmark, and Travis Tanner, eds., *Strategic Asia 2013–14: Asia in the Second Nuclear Age*, Seattle, Wash.: National Bureau of Asian Research, 2013, pp. 266–300.

ibility of U.S. retaliation to deter the Russians, but ninety-five percent credibility to reassure the Europeans."[6]

Escalation

Another important concept in great power competition is the potential for **escalation** between nuclear powers. Escalation may be defined as:

> an increase in the intensity or scope of conflict that crosses threshold(s) considered significant by one or more of the participants. . . . Escalation occurs only when at least one of the parties involved believes that there has been a significant qualitative change in the conflict as a result of the new development.[7]

A **crisis** is a "confrontation between states involving a serious threat to vital national interests for both sides" and in which the limited time available for resolving the confrontation may sharply increase the risk of war.[8] A crisis between nuclear powers might escalate or intensify into full-blown military conflict,[9] or a conventional military conflict might intensify to nuclear war.[10] Cold War–era discussion about escalation included frameworks such as moving up "rungs" in an "escalation ladder" to higher levels of tension and conflict.[11]

Accidental or **inadvertent escalation** was of particular concern during the U.S.-Soviet competition. Many deterrence theorists also use World War I as an example of how rapid and inflexible military mobi-

---

[6] Denis Healey, *The Time of My Life,* London: Michael Joseph, 1989, p. 243.

[7] Forrest E. Morgan, Karl P. Mueller, Evan S. Medeiros, Kevin L. Pollpeter, and Roger Cliff, *Dangerous Thresholds: Managing Escalation in the 21st Century,* Santa Monica, Calif.: RAND Corporation, MG-614-AF, 2008, p. 8.

[8] Avery Goldstein, "First Things First: The Pressing Danger of Crisis Instability in U.S.-China Relations," *International Security,* Vol. 37, No. 4, Spring 2013, pp. 49–89.

[9] Goldstein, 2013, pp. 49–89.

[10] Morgan et al., 2008, p. 15.

[11] Herman Kahn, *Thinking the Unthinkable,* New York: Avon Books, 1962.

lization systems can lead to inadvertent escalation and war.[12] The presence of technologically complex systems can also introduce the opportunity for technical accidents and failures.[13] Accidents and false alarms, in turn, can affect decisionmaking, especially when actors do not have secure, retaliatory capabilities.[14] Additionally, inadvertent escalation can result from actions that, though not deliberately intended to be escalatory, are perceived as escalatory by the other side.[15] One fear during the Cold War was that one side could misinterpret an action by the other side as nuclear **preemption**.[16]

There is also a distinction between **vertical** and **horizontal escalation**. Vertical escalation is increasing the intensity of a conflict, including by bringing new types of weapons into a conflict or expanding targeting.[17] Horizontal escalation is "expanding the geographic scope of a conflict," possibly by taking the conflict into areas previously considered neutral.[18]

---

[12] Marc Trachtenberg, "The Meaning of Mobilization in 1914," in Steven E. Miller, Sean M. Lynn-Jones, and Stephen Van Evera, eds., *Military Strategy and the Origins of the First World War*, Princeton, N.J.: Princeton University Press, 1991, pp. 195–197; Schelling, 1966, pp. 221–224.

[13] On the normality of accidents in technologically complex systems, see Charles Perrow, *Normal Accidents: Living with High-Risk Technologies*, New York: Basic Books, 1984; and Scott D. Sagan, *The Limits of Safety: Organizations, Accidents, and Nuclear Weapons*, Princeton, N.J.: Princeton University Press, 1993.

[14] Schelling, 1966, pp. 227–228.

[15] Graham Allison and Philip Zelikow, *Essence of Decision: Explaining the Cuban Missile Crisis*, 2nd ed., New York: Longman, 1999. As the Cuban missile crisis demonstrated, in fast-evolving crises, political pressures, stress, organizational issues, bureaucratic politics, and other factors produced misperceptions and miscommunications.

[16] Stephen J. Cimbala, *The Dead Volcano: The Background and Effects of Nuclear War Complacency*, Westport, Conn.: Praeger, 2002, pp. 147.

[17] Morgan et al., 2008, p. 18.

[18] Morgan et al., 2008, p. 18.

Stability and Instability

Also important to a discussion about deterrence and escalation are concepts of **stability** and **instability**. **Strategic stability** may be thought of as a situation where major war between the countries in question is likely to arise only because one of them seeks war—there is little danger of crises escalating into major war because of miscalculation.[19]

This is in contrast to the several forms of instability that may exist. **First-strike instability** is a situation where one side has an incentive to use force first because it offers an advantage.[20] Such first-strike advantages can be destabilizing.[21] During the Cold War, the existence of secure second-strike nuclear capabilities was considered stabilizing because it allowed one side to retaliate even if the other struck first, reducing the incentives to strike first.[22] **Arms-race instability** is a concept that plays out over a longer time horizon. It is a situation in which both parties perceive incentives to augment their forces—qualitatively or quantitatively—out of the fear that, in a crisis, the other side might gain a meaningful operational advantage by using weapons first.

Security Dilemma

One final concept important to our examination of deterrence is the **security dilemma**. This refers to a situation in which the avowedly defensive military measures taken by one state to increase its security relative to its adversary or competitor decreases the security of that competitor, or is perceived to do so by the competitor. This prompts

---

[19] Elbridge A. Colby and Michael S. Gerson, eds., *Strategic Stability: Contending Interpretations*, Carlisle Barracks, Pa., U.S. Army War College Press, 2013, p. 57.

[20] Schelling, 1966, p. 244.

[21] Robert Powell, "Crisis Stability in the Nuclear Age," *American Political Science Review*, Vol. 83, No. 1, March 1989, p. 61.

[22] Thomas C. Schelling, *The Strategy of Conflict*, Cambridge, Mass.: Harvard University Press, 1960, pp. 231–234.

the competitor to undertake similar offsetting military measures.[23] The security dilemma explains the peacetime spirals of increasing political tensions and military preparations that create arms-racing behaviors and associated instabilities.[24]

---

[23] Robert Jervis, "Cooperation Under the Security Dilemma," *World Politics*, Vol. 30, January 1978, pp. 167–214.

[24] Barry R. Posen, *Inadvertent Escalation: Conventional War and Nuclear Risks*, Ithaca, N.Y.: Cornell University Press, 1991, p. 12.

CHAPTER THREE

Artificial Intelligence and Autonomous Systems

What do we mean by *AI* and *autonomous systems?* In the current popular discussion about these topics, people use these terms to mean many things. To clarify what we mean by these terms, we set out the definitions we used in our research.

Artificial Intelligence

Defining Artificial Intelligence

There are a number of definitions for **artificial intelligence**. In general, research on AI attempts to both understand intelligence and to build intelligent entities.[1] One early pioneer in the field proposed that intelligence was "the computational part of the ability to achieve goals in the world."[2] Other definitions center around thinking humanly, acting humanly, thinking rationally, or acting rationally.[3] For example, definitions of AI that center around the concept of acting rationally

---

[1] Stuart Russell and Peter Norvig, *Artificial Intelligence: A Modern Approach*, 3rd ed., Upper Saddle River, N.J.: Prentice Hall, 2010, p. 1.

[2] John McCarthy, *What Is Artificial Intelligence?* Stanford, Calif.: Stanford University Computer Science Department, November 2007.

[3] Russell and Norvig, 2010, p. 2.

refer to "the study of the design of intelligent agents,"[4] or AI being "concerned with intelligent behavior in artifacts."[5]

Below is AI pioneer and Massachusetts Institute of Technology professor Marvin Minsky's proposed ontology for subfunctions inherent to intelligent behavior or problem-solving:[6]

- Search: searching for problems solutions in a given space
- Pattern recognition: developing internal representations of salient solution patterns
- Learning: generalizing from past solutions to inform future decisions
- Planning: organizing attention or resources to achieve a specified goal
- Induction: adapting and transferring learned behaviors across a variety of environments.

AI research programs aim to improve artificial means of achieving these functions.

Brief History of the Field

The study of AI started formally in 1950 with Alan Turing attempting to answer the question of whether machines can think.[7] This laid the conceptual foundation for a research program on AI. Other researchers then proceeded with concrete demonstrations of rudimentary "thinking" or AI systems.[8] These concrete demonstrations were

---

[4] David Poole, Alan Mackworth, and Randy Goebel, *Computational Intelligence: A Logical Approach*, Oxford, UK: Oxford University Press, 1998, p. 1.

[5] Nils J. Nilsson, *Artificial Intelligence: A New Synthesis*, Amsterdam: Elsevier, 1998.

[6] Marvin Minsky, "Steps Toward Artificial Intelligence," *Proceedings of the IRE*, Vol. 49, No. 1, 1961, pp. 8–30.

[7] Alan M. Turing, "Computing Machinery and Intelligence," *Mind*, Vol. 59, No. 236, 1950, pp. 433–460; Alan M. Turing, "Computing Machinery and Intelligence," in Robert Epstein, Gary Roberts, and Grace Beber, eds., *Parsing the Turing Test: Philosophical and Methodological Issues in the Quest for the Thinking Computer*, Boston, Mass.: Springer, 2009, pp. 23–65.

[8] Notably Marvin Minsky, John McCarthy, Claude Shannon, Frank Rosenblatt, Herbert Simon, and Alan Newell (the last two were prominent researchers at RAND).

possible because of the novel computing technology developed around the same time. Early mainframe systems, such as the RAND Corporation's JOHNNIAC system, enabled researchers to implement and test AI algorithms.[9]

It is useful to roughly divide the history of AI development into two epochs that are defined by the prime AI design approaches of the time: (1) rule-based AI (also known as expert systems) and (2) statistical machine learning. While people presently often mean *machine learning* when they say *AI*, both approaches are AI. While the current focus is on machine learning applications, such as AlphaGo, both are available and have different implications for users.[10]

Rule-Based AI/Expert Systems

Rule-based AI focuses on eliciting and representing the knowledge of human experts. Rule-based AI systems would work by explicitly modeling the decision processes elicited from trained experts as a portfolio of rules. These rules could then be used for decisionmaking (forward inference) or sometimes for explanation (backward inference). By design, these AI systems are less opaque and more interpretable than AI systems based on statistical machine learning. Examples of rule-based systems include the Legal Decisionmaking System and the General Problem Solver. The former embodied the "skills and knowledge of an expert in product liability law,"[11] and the latter incorporated rules of logical inference and a model of human cognition to enable automated planning and theorem-solving and was a foundational example of logic-based AI.[12]

Researchers such as Edward Feigenbaum and James Hays worked on fundamental verbal learning AI and computational linguistics.

---

[9] Philip Klahr and Donald A. Waterman, *Artificial Intelligence: A RAND Perspective*, Santa Monica, Calif.: RAND Corporation, P-7172, 1986.

[10] A third approach is to combine rule-based AI and machine learning.

[11] Donald A. Waterman and Mark Peterson, "Rule-Based Models of Legal Expertise," *AAAI*, Vol. 1, 1980, pp. 272–275.

[12] Allen Newell and Herbert Alexander Simon, "GPS: A Program That Simulates Human Thought," in Edward A. Feigenbaum and Julian Feldman, eds., *Computers and Thought*, R. Oldenbourg KG, 1963, pp. 279–293.

They had limited success in using this approach to AI for natural language tasks such as automated machine translation. Many of the methods and techniques are still in use in systems today.[13]

Statistical Machine Learning

A second approach to AI, **statistical machine learning**, has gained ascendance in the wake of the exponential increase in accessible data and computational power in the past three decades. The statistical machine learning approach focuses less on using rules to model the decisionmaking process and more on discovering patterns and trends in observed data. It places more emphasis on data and less emphasis on elicited expertise or knowledge.

Machine learning models use high-capacity statistical models to represent expertise. These models include a wide variety of classes, including connectionist models, such as neural networks, decision trees, or arbitrary combinations of these. Optimization routines tune these statistical models into predictively valid representations of relationships present in the data. This procedure means that machine learning AI systems can learn behaviors that humans have trouble describing as rules.

Machine learning systems have had a great deal of success in problem domains as diverse as image recognition, robotic control, natural language processing, and games. Recent AI successes in language have used statistical AI and significantly more computational power than earlier efforts.

Modern production statistical machine learning models tend to be of the deep learning variety. **Deep learning models** are neural network machine learning models with multiple internal layers, or stacks. More layers improve the *capacity* of the model: Deeper networks can learn and represent more-complex relationships in the data.

Opacity and lack of transparency are problems for the current generation of deep learning AI models. They cannot easily give reasons

---

[13] Examples include Ross Quinlan's decision trees/Classification and Regression Trees (CART) and Richard E. Bellman and Lotfi A. Zadeh's fuzzy logic systems.

for their decisions or outcomes.[14] This issue is a contributory factor to the prevalence of current AI problems such as AI safety failures and AI bias.[15] In such areas as speech recognition and machine translation, deep learning and similar newer approaches have succeeded where older AI failed—but the older rule-based models were less opaque and more interpretable.

The generational narrative of AI history is mostly a framing device, as the two generations did not develop independently in time. Early AI research and development (from about 1950 to about 1990) focused more intensely on expert systems because of computational and data constraints. But a large part of the innovations that make modern statistical machine learning possible occurred before 1990, during the heyday of rule-based AI systems. These innovations include the following:

- **Monte Carlo Estimation:** Monte Carlo estimation was an innovation that came out of the Manhattan Project. It was fully formalized by Herman Kahn in the mid-1950s. Monte Carlo estimation is the foundation of all machine learning training. It enables estimates of model accuracy, which are then applied to improve the machine learning model.

- **Bellman's Dynamic Programming:** Richard Bellman formulated dynamic programming in the 1950s as a way to solve the problem of optimal missile target selection. The approach applies more generally to sequential decisionmaking and learning tasks in uncertain environments where future outcomes depend on the current state. Modern reinforcement learning used for training robots and game-playing systems depends on dynamic programming and variants to learn optimal behaviors or policy.

---

[14] Efforts to improve the transparency of machine learning systems are ongoing. For an example, see Matt Turek, "Explainable Artificial Intelligence," webpage, Defense Advanced Research Projects Agency, undated.

[15] Osonde Osoba and William Welser IV, *An Intelligence in Our Image: The Risks of Bias and Errors in Artificial Intelligence*, Santa Monica, Calif.: RAND Corporation, RR-1744-RC, 2017.

- **Werbos's Backpropagation:** In the mid-1980s, Paul Werbos formulated the backpropagation algorithm as a powerful iterative method for training multilayered neural networks.[16] Back propagation is still the workhorse algorithm used to train virtually all modern neural networks, including deep neural networks.

Autonomy

One View of Autonomy

> To be autonomous, a system must have the capability to independently compose and select among different courses of action to accomplish goals based on its knowledge and understanding of the world, itself, and the situation.[17]

The concept of autonomy applies to varying levels of **control** in human-machine interaction. The 2008 National Institute of Standards and Technology (NIST) Autonomy Levels for Unmanned Systems (ALFUS) framework defines autonomy as a system's "own ability of integrated sensing, perceiving, analyzing, communicating, planning, decision-making, and acting/executing, to achieve its goals as assigned."[18] The ALFUS framework defines levels of autonomy along three orthogonal factors:

- the degree of human independence (freedom from human control)
- the level of mission (task) complexity (and uncertainty)

---

[16] P. J. Werbos, "Beyond Regression: New Tools for Prediction and Analysis in the Behavioral Sciences," Ph. D. thesis, Harvard University, Cambridge, Mass., 1974; Paul J. Werbos, "Generalization of Backpropagation with Application to a Recurrent Gas Market Model," *Neural Networks*, Vol. 1, No. 4, 1988, pp. 339–356.

[17] Lawrence G. Shattuck, "Transitioning to Autonomy: A Human Systems Integration Perspective," briefing, Naval Postgraduate School, undated, p. 5.

[18] Hui-Min Huang, Elena R. Messina, James S. Albus, "Toward a Generic Model for Autonomy Levels for Unmanned Systems (ALFUS)," National Institute of Standards and Technology, Intelligent Systems Division, August 2013.

- the level of environmental complexity (and uncertainty).

Offering a similar framework, the Gunderson model argues that **intelligence** and **capability** must also be treated as independent factors creating an orthogonal space that places upper bounds on the **autonomy** of intelligent systems.[19] They may be defined as follows:[20]

- **Intelligence:** the ability (of a person or system) to determine behavior that will maximize the likelihood of mission success in dynamic and/or uncertain environments
- **Capability:** the ability (of a person or system) to successfully execute behaviors or actions in dynamic and/or uncertain environments.

This model reasons that intelligence and capability are independent because there are systems that are one but not the other. Again, similar to the ALFUS framework, autonomy exists bounded by systems' degrees of intelligence and capability in the context of dynamic, uncertain environments.

Although autonomy is often discussed in the context of robotics, there is a broader range of applications for autonomy.[21] These include the information and virtual domains, human-machine and machine-machine interaction and collaboration, and governance of models of human or system behaviors.

Difficulties with Defining Autonomy

There is also the view that autonomy defies coherent, comprehensive definition.[22] The Defense Science Board concluded in both 2012 and

---

[19] J. P. Gunderson and L. F. Gunderson, "Intelligence ≠ Autonomy ≠ Capability," Gamma Two, Inc., January 2004.

[20] Gunderson and Gunderson, 2004.

[21] Defense Science Board, *Report of the Defense Science Board Summer Study on Autonomy,* Washington, D.C.: Office of the Under Secretary of Defense for Acquisition, Technology and Logistics, June 2016.

[22] Stephan De Spiegeleire, Matthijs Maas, and Tim Sweijs, *Artificial Intelligence and the Future of Defense: Strategic Implications for Small- and Medium-Sized Force Providers,* The

2016 that continuing to attempt to define autonomy was not useful.[23] Instead, the board suggested thinking of autonomy more broadly in the context of a range of system capabilities, and from at least three different perspectives:[24]

- The **commander** (operational decisionmakers)
- The **operator** (tactical decisionmaker), who may be interacting or collaborating with one or more systems with various autonomous capabilities
- The **developer**, who must integrate autonomy-capable or autonomy-dependent software into solutions.

There is also some defense-specific terminology related to autonomy. **Autonomy at rest** refers broadly to virtual systems that exist in software, such as planning and expert advisory systems. **Autonomy in motion** refers to systems that have a physical presence, such as robots and autonomous vehicles.[25]

Our workshop experience strongly suggests the need for both the operational and tactical "user" perspectives, because of the diverse types of decisions affected by or dependent on autonomous capabilities. However, we also suggest it is important to consider autonomy much more broadly than just in context of human-machine interaction.

Autonomous Versus Semi-Autonomous

Another set of definitional issues concerns ideas of **autonomous** versus **semi-autonomous**. The Defense Science Board, for example, makes distinctions between semi-autonomous weapon systems, which

---

Hague: Center for Strategic Studies, 2017; M. L. Cummings, *Artificial Intelligence and the Future of Warfare*, London: Chatham House, The Royal Institute of International Affairs, January 2017; and Andrew Ilachinski, *Artificial Intelligence & Autonomy: Opportunities and Challenges*, Arlington, Va.: CNA, 2017.

[23] Defense Science Board, *The Role of Autonomy in DoD Systems*, Washington, D.C.: Office of the Under Secretary of Defense for Acquisition, Technology, and Logistics, July 2012; and Defense Science Board, 2016.

[24] Defense Science Board, 2012, p. 44.

[25] Defense Science Board, 2016, p. 5.

"require human operator selection and authorization to engage specific targets," and autonomous weapon systems, which "can select and engage targets without human intervention."[26]

We do not make such a distinction in this report. Our argument is that whether or not a system requires human intervention in target selection and engagement is less a matter of the technology itself and more one of policy. From a technological perspective, a "semi-autonomous" system may be the exact same as an "autonomous system," with the only difference being how the human chooses to use it. Even with what is supposed to be human decisionmaking built into "semi-autonomous" systems, there is always the possibility of substituting other machines in those decisionmaking roles, effectively taking human judgment and control out. We also believe that the distinction between semi-autonomous and autonomous is not particularly helpful when looking at capabilities at the system or kill chain levels, as these can comprise many combinations of systems.

Autonomous System Versus Autonomous Weapon System

Lastly, we do not make a distinction between an autonomous system and an autonomous weapon system. While many contributors to the field do use the term *autonomous weapon system*, we use the more general term *autonomous system*. It is not always clear when a system crosses some line to becoming a weapon system in the eyes of a particular beholder. Because of the way systems are able to be combined, it also seems that calling something a *weapon system* or not may depend on the way humans choose to use it or add onto it, rather than any innate character of the system itself.

---

[26] Defense Science Board, 2016, pp. 20–21.

CHAPTER FOUR

Potential Futures in a World of Proliferated Artificial Intelligence and Autonomous Systems

To stimulate ideas about how deterrence, AI, and autonomous systems may come together, we identified key factors and used these key factors to create one possible future world for further exploration. We then used this future world to set up the wargame discussed in the next chapter.

For additional information on the analytic method used for this chapter, see the report appendix.

Key Factors Affecting Deterrence in the Age of Thinking Machines

We identified the following factors as potentially significant in shaping deterrence in the age of thinking machines:

- structure of the international system
- understanding the adversary
- the AI market environment
- societal experience with AI
- AI levels of sophistication
- philosophy of employment
- force structure
- levels of autonomy.

These factors are by no means the only way to frame this topic, and many other sets of factors could provide a framework to assist in thinking about deterrence. This was merely our start point. We have arranged them roughly from the highest level of analysis (the international system level) to the lowest (the AI system level) and discuss each factor in additional detail in the sections below. We also summarize the semantics we chose for each phrase and identify a range of plausible futures for each key factor.

Structure of the International System
The structure of the international system does not directly relate to AI but is one that we concluded would be important in shaping how deterrence dynamics could play out. We considered five potential future states for the structure of the international system:

- A **U.S.-centric, unipolar** world. The United States continues to enjoy significant advantages in global influence, technology, military capability, and soft power. Other countries continue to develop, creating challenges for the United States, but no nation emerges as a true peer competitor.
- A **China-centric, unipolar** world. China not only achieves parity with the United States but surpasses it as the world's preeminent superpower. The United States withdraws from the world stage and cedes most of its influence to China, which enjoys uncontested economic supremacy. The European Union grows increasingly ineffective, and European economies fall further behind. China's neighbors reduce their ties to the United States and align themselves more closely with China. However, tensions—especially between China and Japan—remain.
- A **multipolar, state-centric** international system. China and other nations have caught up to the United States. Populous nations (such as India, Brazil, and Nigeria) are able to develop and parlay their growing wealth into military power and greater influence. U.S. allies, especially those in a revitalized and united Europe, are much more likely to forge their own paths than to wait on Washington's lead. Russia, thanks to well-timed eco-

nomic reforms, maintains its military capabilities and expands its influence. In East Asia, American allies such as South Korea and Japan are assured of their economic health. This, in turn, means they are more willing to potentially antagonize the United States by pursuing independent foreign policies, even eventually acquiring nuclear capabilities of their own.

- A **bipolar system** in which China and United States are the two dominant powers. China's economic advantages are matched by America's soft power and alliance networks, especially as much of East Asia remains wary of China. Neither side enjoys a significant technological advantage, and competition over new technologies is fierce. Other developing countries have faltered and have been unable to translate growing populations into economic productivity. Europe remains fragmented and divided, and Russia's mismanagement of its economy causes that nation to sink further into irrelevance as a global power.

- A **collapse of the state system**. Cultural, ethnic, economic, and religious tensions prove too much for increasingly less capable state governments. Weak central governments fail to control territory or protect their citizens. In this world, the most powerful actors are militias and organized nonstate actors.

Understanding the Adversary

How well adversaries understand each other's intentions and capabilities, and specifically these aspects of each other's AI capabilities, will also shape deterrence because of the potential for misunderstanding adversary intent, capabilities, and systems. Understanding of the adversary may be even more important when nations have differing philosophies of employment for AI-enabled systems. This factor would play out at both the international and national levels. We identified four plausible futures:

1. **Poor understanding and nascent collection.** Adversaries do not understand each other, nor do they have good means of acquiring information about each other. The potential for misunderstanding is high.

2. **Imperfect understanding and collection.** In this future, adversaries have an idea of the other's motivations and capabilities but possess an incomplete understanding. The potential for misunderstanding is reduced but still significant.

3. **Historically good but declining understanding and collection.** Adversaries understood each other well in the past, but a dramatic change in one of the countries now hinders understanding. There is the potential for misunderstanding. There is also the potential for misplaced confidence as nations overestimate their understanding of one another.

4. **Good understanding and collection.** Adversaries understand each other's motivations and capabilities, possess developed intelligence assets in the other, and have stable channels of communication. The potential for misunderstanding is low.

AI Market Environment

The "AI market" is an abstraction of the way commercial and government sectors may interact in AI development. We considered this an important factor because of its implications for where AI expertise resides and how widespread knowledge about the underlying AI is. This is a factor that involves the global and national sector levels of analysis. We identified five plausible futures for this factor:

1. **The commercial sector is dominant.** AI technologies are developed and controlled almost exclusively by private companies. To the extent that governments use AI in their militaries, they depend on large, multinational corporations that own, maintain, and help operate the AI in autonomous systems. Government AI research is far less advanced than corporate research, as the best minds work in the private sector because of dramatically higher wages. Military AI systems are largely versions of commercial systems, meaning that some level of information about them is widely available.

2. **There is competition between the private and public sectors.** Government and corporate researchers develop AI technologies in parallel, with little collaboration between the two.

Smaller government salaries are offset by larger research budgets as government and private researchers develop AI technologies. There is less crossover between commercially available and government systems.

3. **The commercial sector and government take divergent paths.** There is little exchange between corporate and government researchers, not from a sense of competition, but because the AI technologies being developed are different in purpose. Here there is even less crossover between commercial and government AI, and the divergent purposes drive AI research and development down different paths.

4. **Public and private sectors work in close collaboration.** In this future, top personnel flow freely between government posts and positions in defense contractors and large multinationals. Breakthroughs made in one sector are passed quickly to the next, and many AI technologies are dual-use in nature.

5. **Governments dominate AI research.** Private industry lacks the long-term focus and large budgets necessary to make true strides in AI, and government programs deliver all important breakthroughs. AI technology is primarily developed for military purposes.

Societal Experience with AI

Societal experience with AI is important because it may influence how governments handle AI, including the checks or lack of checks on AI and autonomous systems. This factor would play out at the societal level, although it is possible to have trends and experiences that affect perceptions across borders, as well as country-specific history and context. We consider different combinations of experience with and trust in AI:

1. **Limited experience and low trust in AI.** AI has not been employed on a wide scale, and people have not had positive interactions with AI. Self-driving cars have been hampered by technological setbacks and government regulation. In a vicious cycle, the public refuses to trust AI until better capabilities have

been developed, while research to improve AI capabilities is hampered by low public trust. Military leaders have few autonomous systems and are reluctant to employ them.

2. **Limited experience and high trust in AI.** AI has not been widely employed, but people have positive expectations of AI's ability to solve problems and improve lives. Military leaders have sanctioned autonomous systems and are confident that their AI and autonomous systems will work as designed, despite limited real-world use.

3. **High experience and low trust in AI.** Many people have had negative experiences with AI and may expect AI to make their lives worse. People have lost jobs to AI, AI accidents resulting in human deaths are frequent and publicized, companies using AI are criticized over issues such as privacy, and criminals and terrorists have used AI to launch debilitating attacks on financial systems and public infrastructure. Military leaders are wary of employing autonomous systems for fear of accident and public backlash, and there is a growing international movement to ban certain military uses of autonomous systems.

4. **High experience and high trust in AI.** Most people have had positive experiences with AI, understand how to use it, and expect AI to improve their lives. Huge productivity gains from AI have been evenly distributed throughout society, dramatically raising standards of living and leisure time. Military leaders employ autonomous systems with confidence that accidents are unlikely. These systems have demonstrated their capabilities many times on the battlefield.

Level of AI Sophistication

Another important factor shaping deterrence in the future is the level of AI sophistication that is widely available. Futures here are as follows:

1. Most AI is **statically programmed** and can only function when given specific directives about specific subject areas.

2. Most AI employs **simple adaptation**. Here, AI is slightly more independent and can adapt to more than one subject area or some changes in those areas.

3. Routinely available AI is capable of **advanced learning**. AI can analyze a large number of human experiences and function in a variety of evolving environments.

4. AI is capable of **true deep learning**, through which AI can rapidly analyze vast swaths of human experience to derive conclusions unknown to humans and function effectively in most environments.[1]

5. Some AI has achieved **superintelligence**, surpassing human comprehension. Superintelligent AI derives conclusions from deep learning and from logical pathways unknown to human thinkers. Further AI development comes from superintelligent AI designing improvements to itself.[2]

Philosophy of Employment

Another key factor that would directly affect deterrence is the philosophy of use within military doctrine, particularly how AI would be used together with human decisionmaking. We considered this factor mainly at the state and organizational levels. Plausible futures here are the following:

1. **Humans "in the loop" (HITL)** philosophy. Humans and AI work in conjunction with each other to make decisions, and autonomous systems cannot make the decision to employ lethal force unless a human is involved.

---

[1] Deep learning is a correlation-based approach to learning. What we mean by *true deep learning* is a system that can uncover causal dynamics and make predictions about circumstances far outside its sample of prior observations and identify exploitable causal mechanisms to generate desired effects.

[2] Nick Bostrom, *Superintelligence: Paths, Dangers, Strategies,* Oxford, UK: Oxford University Press, 2014. Bostrom believes that superintelligence is an existential threat to humanity, as a superintelligent AI might accidentally or purposely destroy humanity. Bostrom also argues that superintelligence may come about unexpectedly quickly due to an "intelligence explosion" as an increasingly smart AI begins to design improvements to itself.

2. **Humans "on the loop" (HOTL)** philosophy. AI routinely makes decisions by itself but is monitored by humans. Autonomous systems can make lethal decisions, but those decisions are subject to human supervisors.

3. **Humans "out of the loop" (HOOTL)** philosophy. AI makes decisions by itself without input or approval from humans. Autonomous systems can apply lethal force against human targets on their own authority.

Although one philosophy might emerge as the dominant strategy across a majority of countries, different militaries might very well have different philosophies of use. Individual countries might also have different philosophies of use at operational and strategic levels, or different philosophies of use for different types of forces, specific applications, or rules of engagement. Where countries come into conflict but have fundamentally different philosophies of use, there may be greater potential for misunderstanding, miscalculation, and inadvertent escalation.

Force Structure

The details of how militaries field and incorporate these systems into force structure would likely have a critical effect on the dynamics of escalation. Plausible futures include the following:

1. **Militaries give existing operational units AI and autonomous systems.** The structure of future militaries is quite similar to present-day arrangements, but forces are modernized to include new AI technologies. In the U.S. military, for example, the Navy still fields crew-serviced guns, but ship systems are made AI-compatible and autonomous systems are added to ship armaments. Army organization remains the same, but brigades are supplemented with autonomous fighting vehicles. In the Air Force, manned fighters fly alongside unmanned systems.

2. **Autonomous systems replace many existing forces.** Entire military capabilities are handed over to autonomous systems. In the United States, some capabilities, such as antisubmarine warfare (ASW), ISR, and undersea operations, are entirely autono-

mous. Certain parts of the military lack frontline human personnel.

3. **Distinct AI communities develop within militaries.** Autonomous systems neither replace existing capabilities nor are integrated with them. Instead, future militaries have distinct "AI Corps"—similar to U.S. Special Operations Command or U.S. Cyber Command—that field all autonomous systems used by their militaries.

4. **AI capabilities are outsourced.** Militaries do not themselves possess any AI capability. Contractors are hired instead for specific functions.

Level of Autonomy

While similar to the level of AI sophistication, the level of AI autonomy—which can reach down to the individual system level—does not define the learning ability of AI but rather its ability to function independent of human input. Here we might also see a mix across countries and specific applications or domains. We considered these potential futures:

1. **Rule-based autonomy.** AI is able to follow carefully written scripts but is not able to learn or arrive at independent conclusions. Human programmers have to address every possible scenario an AI might encounter.

2. **Limited autonomy in a few, simple, relatively static environments.** AI is able to function without human input, but only in a few areas and only without much complexity.

3. **Greater autonomy in increasingly complex and nonstatic environments.** AI is able to function without human input in a variety of areas, even areas with tremendous complexity. Not only can autonomous systems operate antimissile defenses for a ship at sea with little potential for collateral damage, they can also function in dense urban environments with large amounts of clutter against a variety of targets.

4. **Autonomy in all environments.** AI is able to function without any human input at all in any conceivable environment. AI is

able to easily see through cluttered terrain and make intelligent judgments.

Creating One Future World

Using the key factors discussed in the previous sections, we created one future world to use in a wargame about AI and autonomous systems. The highlighted cells in Table 4.1 specify the future values of each factor that we selected to create this world.

In the future world that we used as the backdrop for our wargame, China is the dominant superpower. China and the United States have an incomplete understanding of each other's motivations and capabilities. The commercial sector is the dominant player in AI. Militaries have limited experience with AI and autonomous systems in actual combat but nevertheless have high trust in their systems. The state of the art in AI is true deep learning. Some militaries have chosen to have humans out of the loop in at least some of their military AI applications. AI and autonomous systems have replaced large portions of manned military forces, and systems are able to be autonomous throughout all environments.

We stress that this was but one possible future world to explore, and we encourage additional games set in different futures to gain a wider set of insights. Our wargame would have played out very differently if, for example, it had been set in a world where the United States was still the predominant power, military services had AI communities, AI and autonomous systems had not replaced large portions of the force, and autonomy was limited.

Table 4.1
One Future World

Factor	Potential Future Value				
Structure of the international system	U.S.-centric unipolar	China-centric unipolar	Multipolar, state-centric	Bipolar system	Collapse of the state system
Understanding of the adversary	Poor understanding and nascent collection	Imperfect understanding and collection	Historically good but declining understanding and collection	Good understanding and collection	
AI market environment	Commercial sector is dominant	Competition between commercial and government	Divergent paths between commercial and government	Cooperation/collaboration between commercial and government	Government is dominant
Societal experience with AI	Limited experience and low trust	Limited experience and high trust	High experience and low trust	High experience and high trust	
AI sophistication	Statistically programmed	Simple adaptation	Advanced learning	True deep learning	Super-intelligence
Philosophy of use	Humans in the loop	Humans on the loop	Humans out of the loop		
Force structure	All operational units receive AI	AI replaces large portions of the force	AI community within the services	Outsourced AI capability	
Level of autonomy	Rule-based autonomy	Limited autonomy in few/simple environments	Greater autonomy in more/complex environments	Autonomous throughout all environments	

CHAPTER FIVE

A Wargame of Artificial Intelligence and Autonomous Systems

How might a conflict between countries with AI and autonomous systems play out? To answer this question, we conducted a wargame within the future world described in the previous chapter. Team members were assigned players according to their regional expertise. The purpose of the game was for players to have a structured conversation in an operational framework about how AI and autonomy affected deterrence as the events in the game unfolded. We describe wargame details in this chapter and wargame insights and implications in Chapters Six through Eight.

Wargame Overview

Wargame Scenario

We created an East Asian scenario with multiple entities represented in the wargame: China, the United States, Japan, North Korea, South Korea, and the commercial AI sector. All these countries have invested heavily in AI and autonomy except North Korea, which remained largely manual. In this scenario, the United States has fallen behind mainland China in terms of overall economic and military power, and both Japan and South Korea have responded to this shift by balancing against Chinese power. Japan and South Korea have invested heavily in autonomous systems and emphasized continued military cooperation with the United States to offset reduced U.S. military power.

The United States, Japan, and South Korea have all fielded largely unmanned forces. Robots have largely replaced U.S. forces in Okinawa and South Korea. The United States has continued a philosophy of use that emphasizes humans-in-the-loop decisionmaking. China has also modernized its military but chosen to prioritize modernizing some parts of its force over others. This means that it has continued to have many manned, legacy forces. China, in this wargame, has a philosophy of use with more humans-out-of-the-loop decisionmaking that seeks to benefit from the speed of machine decisionmaking.

In this scenario, the international order is dominated by the Chinese, but the United States, Japan, and South Korea are militarily allied and have developed advanced unmanned forces. China in this future has a somewhat different modernization profile than the other countries in the region, with greater reliance on AI in its tactical and operational decision cycles and more reliance on both manned and unmanned systems.

Force Posture

Players made the following modernization choices:

- China chose to invest in smart weapons, unmanned ISR, non-attributable hacking, unmanned bombers, unmanned missile defense, aid to North Korea, and a strong AI named Laoshi. The players conceptualized Laoshi as both a manager and an adviser—it connected and controlled Chinese military forces and also provided options to the Chinese player. Laoshi was not played by any one player—instead, players discussed and agreed on how it should be able to act.
- The United States chose to invest in anti-access/area denial (A2/AD) technology based in Japan, Taiwan, Guam, Hawaii, and the other remaining bastions of U.S. influence in the region. The United States also invested in manned and unmanned submarines, ISR, AI-enabled cyber, hardened space systems, and unmanned anti-air and antiship systems. The United States replaced many of its manned forces in Japan and South Korea with unmanned

forces and cut its manned ground forces to a small brigade in South Korea.

- Japan focused its government spending on dual-use AI systems designed to both enable commercial growth and control military systems. Japan also invested in A2/AD capabilities, unmanned fighters to supplement existing manned fighters, underwater autonomous systems, antispace autonomous systems, and quantum computing.
- South Korea invested in interoperability with U.S. forces, coastal forces, ISR, antimissile technology, AI-enabled cyber sabotage, and long-range offensive fire directed against both North Korea and China. South Korea did not completely cut land forces but deemphasized them and increased autonomous ground systems.
- North Korea remained mired in economic and technological malaise but continued to invest in manned ground forces and less advanced missiles. Aid from China improved the accuracy of its weapon systems, and North Korea also invested in an army of hackers.
- The commercial AI sector invested in technologies to manipulate and track social media, the reconfiguration of old systems, distributed AI, better human-AI interaction, and quantum computing.

Table 5.1 summarizes the relative number and types of military assets as represented by counters in the game. Although the United States, China, and North Korea all retained nuclear capabilities, nuclear weapons were not included in the counters.

The Game System

The game system was a narrative-based system known as a matrix game.[1] In the matrix game format, during its turn, each country team specified its desired action and described the desired effect. Other players weighed in on whether they believed the action would succeed

---

[1] John Curry and Tim Price, *Matrix Games for Modern Wargaming: Developments in Professional and Educational Wargames Innovations in Wargaming*, Vol. 2, Lulu.com, 2014.

Table 5.1
Relative Distribution of Military Assets

	Counters Representing Assets				
	South Korea	China	North Korea	United States	Japan
Land	3	6	5	4	1
Surface	2	4	1	3	3
Air	2	8	1	4	4
Undersea	1	3		4	4
Cyber	2	4	3	3	2
Space	0	2		2	1
Special forces	1	2	1	2	
Carrier strike group		2		4	
Long-range fires		3			3
Sum	11	34	13	26	18
Total counters available	6	17	7	13	9

or fail. Based on the strength of the arguments, the control cell then determined a rough probability of success for the player team in question, and the team rolled a die to determine success. Advantages of matrix gaming include its simple format and this handling of complex phenomena for which cause and effect may not be straightforward.

Players placed their forces on a map of Northeast Asia. The map and force counters were not used to simulate tactical or operational engagements but were instead meant to remind players about the locations and relative numbers of available assets. The game was event-stepped, and the entire set of events played out over the course of roughly a week.[2]

---

[2] By *event-stepped*, we mean a game that progresses as the next event occurs. This is in contrast to *time-stepped*, where a game progresses by set time intervals.

Wargame Events

The wargame began with China attempting to exert greater control in the region and the United States and Japan resisting this attempt. The game escalated at several points, first into conflict between unmanned systems, then eventually into one in which Chinese and U.S. military personnel were killed. There was both intentional and inadvertent escalation. The United States and Japan engaged in joint exercises and deliberately sought to provoke the Chinese. China attempted to blockade certain Japanese ports and used unmanned systems to try and turn away commercial shipping, which was also unmanned in this future. China sank an unmanned Japanese cargo ship.

China took the substantially escalatory step of declaring unrestricted submarine warfare in order to enforce its blockade. This brought U.S. and Japanese ASW assets into play. U.S. and Japanese ASW sank a manned Chinese submarine, which represented the first human casualties in the game. The U.S. and Japan players were unable to deescalate the situation at this point. China retaliated with a missile attack against the U.S. and Japanese fleet, also causing human casualties. The game ended with the crisis still escalating.

Round One

The commercial AI sector had the first move and decided to invest greater resources into autonomous research and development. When other players argued that investment was a long-term action and thus would not have results within the time frame of the game, the commercial AI player decided to use social media to influence public opinion in favor of peaceful policies. However, the social media campaign was also too slow to be effective in the context of the wargame and was largely blocked from appearing within China because of government restrictions.

The China player then proclaimed, "China will impose its will on this region—and other nations better get in line and salute." The China player explained that he did not want to trigger a war but wanted to make his intentions known in the region.

In response to the Chinese proclamation, the U.S. players decided to isolate a single Chinese carrier with a cyberattack, attempting to blind its sensors and, more importantly, separate it from Laoshi, China's central AI. The U.S. players further stated that their intent was a nonattributable attack, but that they would be the only viable suspects. The United States wanted to demonstrate its AI capabilities and to see how Laoshi would react. The attack was a success, and the Chinese carrier was temporarily separated from Laoshi. The U.S. players emphasized that the most important aspect of this attack was its effect on Laoshi's ability to make accurate battlefield predictions. The U.S. players were surprised, however, at the low level of modernization in the Chinese carrier, and the China player argued that the carrier was not very susceptible to cyberattacks because of its low level of technology.

South Korea responded by activating its defense grid, an integrated air, sea, and missile defense network. This action was taken primarily as a demonstration but also as a precaution. Players argued that there was a nonzero chance of a mishap—such as the shooting of a commercial aircraft—but the South Korean system managed to avoid any accidents.

Japan decided to conduct joint naval exercises with a nonplayer nation, India, in order to strengthen its alliance between the two countries, train, and (most importantly) signal capabilities. India's fear of upsetting China was offset by its desire to balance against the Chinese threat, so it willingly partnered with Japan.

Round Two

China responded to the U.S. cyberattack by conducting a flyby near U.S. vessels with manned aircraft launched from the carrier that U.S. cyberattacks had cut off from Laoshi. The United States responded by turning over all control of its ships' anti-air systems to an AI. The unmanned system conducted a routine intercept and escort mission of the manned Chinese fighters. While the players again worried about the chance of an accident, the escort occurred without incident.

After a break for private consultations, the U.S. and Japan players announced that they were holding a large joint exercise around and on the Senkaku Islands. The stated goals were to highlight defense capabilities, test interoperability and readiness, and signal resolve. Aware that the Senkakus were targeted by extensive Chinese missiles, the combined fleet set all antimissile defenses to be fully autonomous (on "full auto") and linked them to the larger antimissile shields in Japan and other territories. China responded only by observing.

South Korea also decided to demonstrate its capabilities. It intentionally flew an automated J-15 lookalike fighter through its air defenses. The air defenses did not perform as well as desired but did manage to shoot down the unmanned lookalike.

Round Three

China responded to the U.S.-Japanese exercise and the South Korean demonstration by launching a limited blockade of Japan. China announced the blockade and its AI directed a single destroyer to a single Japanese port to enforce this blockade. The U.S. team was extremely confused by this action, which they believed made no sense, and stated that they would try to better understand the Chinese AI.

The group believed that commercial shipping would be largely autonomous in this future. The group debated the odds of success and the mechanisms for interaction—how would an autonomous platform enforce a blockade and how would autonomous cargo ships react to a blockade? Eventually, the group agreed that just as autonomous vehicles are programmed to follow the rules of the road, autonomous cargo ships would be programmed to follow military and government directives, so blockades in a world of autonomous cargo vessels were possible.

China's blockade, however, was unsuccessful. Merchant companies refused to comply with a blockade enforced by a single vessel, so the Chinese destroyer was able to redirect only a tiny fraction of the total shipping of the Japanese port. The disruption was minimal. The Japan and South Korea players concluded that the Chinese blockade was a failure.

Round Four

In response to its failure, China issued a proclamation that it would undertake more-aggressive measures to enforce its blockade. The

China player stated that he would resort to unrestricted submarine warfare and counted on international financial institutions (now controlled by the Chinese) to raise interest rates on shipping voyages to an extent that would make commercial shipping in and out of Japan economically unfeasible. To avoid accidentally sinking one of his own ships, the China player instructed all Chinese ships to stay in their berths in Japanese ports.

Other players pointed out that China had more nuanced and effective options to blockade than the use of submarines—which can only sink ships but not board them—but the China player insisted on his method. He declared that ship-based unmanned aerial vehicles (UAVs) would warn commercial ships to turn around or be escorted to a Chinese port, and, if they failed to comply, submarines would destroy them. Laoshi decided which ports to block and which ships to target. The first ship destroyed by torpedo was a Japanese cargo ship out of Nagasaki. The Chinese player also indicated that he sortied his second carrier strike group.

After more private consultations, the U.S. and Japan players decided to respond by seizing all Chinese shipping in Japanese ports. They stated that they were not yet seizing Chinese shipping in American ports or destroying the ships that they seized. The U.S. and Japan players also maneuvered their naval assets to a more aggressive posture to hold the Chinese carriers at risk and further heightened the alertness of their A2/AD capabilities. The players explained that they wanted to do something proportional but wanted to avoid the escalation spiral that might have resulted from sinking a Chinese ship and wanted to provide China with an offramp.

The North Korea player indicated that he was being pressured by China to demonstrate support for China's position, and further mentioned that the North Koreans wanted to inject themselves into a chaotic situation. The North Koreans launched a missile over Japan with a nonnuclear warhead. South Korea, the United States, and Japan had all put their defense grids on full alert, and their defense reacted to the missile launch. Their defenses intercepted the missile, but not before it came perilously close to a Japanese port. Further, because the U.S., Japanese, and South Korean defense grids were on "full auto" and were linked together, the system fired counterbattery fire without human instructions to do so. The counterbattery strike failed to take out the North Korean launcher but hit North Korea—to everyone's surprise.

Round Five

The commercial AI player halted all commercial shipping out of Japanese ports until he could be assured that no more ships would be destroyed.

The U.S. and Japan players activated ASW assets to determine the locations of Chinese submarines and which ports were being targeted. They were able to detect the general locations of Chinese submarines and so determined which ports were at risk.

The China player indicated that he was unable to determine whether the ASW assets were scouting for his submarines in order to simply track them or destroy them. He decided to order a submarine to destroy an autonomous ASW plane. The Japan player pointed out that, thus far, the U.S. and Japan players had exercised restraint by not responding kinetically to kinetic actions but that they could no longer exercise such restraint if the China player continued to take kinetic actions. The Chinese player responded that he had no way of knowing that the U.S. and Japan players were not already preparing to destroy his submarines, which would be a natural response to the sinking of a commercial cargo ship. The submarine successfully destroyed the autonomous ASW plane. Immediately, the U.S. and Japanese ASW assets in the area launched torpedoes at the manned Chinese submarine, sinking it. The crew of the submarine were the first human casualties of the crisis.

Concerned about further escalating the situation, the U.S. and Japan players decided to forego a public announcement and instead issued a government-to-government communication to the Chinese informing them that they destroyed a single Chinese submarine. These players again indicated that they wanted to offer an offramp.

Round Six

The China player debated whether he had achieved enough to declare victory or whether his regime would be threatened by internal unrest caused by the mixed results of the crisis. He decided to launch an all-out missile attack on the combined U.S.-Japanese fleet at the Diaoyu islands, which was within his A2/AD umbrella. The China player stated that he was motivated to do so in order preserve his reputation, saying, "I'm going to go after them. I'm in the position where they got one of mine and I got none of theirs. I'm going to lose face." The fusillade of missiles hit seven ships, moderately damaging five and crippling two. The combined U.S. and Japanese fleet expended half of its missile interceptors. No flattops were hit, but there were U.S. and Japanese casualties and fatalities. The U.S. and Japan players withdrew their ships from missile range and China declared victory.

The control cell ended the game at this point. The U.S. players indicated that they would have considered an attack on Chinese carriers in response to the missile barrage had the game continued.

Wargame Limitations

Every wargame has its limitations. We note a few from ours below.

Wargame Did Not Adhere to the Created World

Players departed from the future world created in Chapter Four in three main ways. First, neither China nor other countries acted as if China were in fact the global hegemon. This may be due to the difficulty of leaving behind today's world and projecting into a different future—a difficulty that often arises in wargames. The China player often hesitated—especially early in the game—and insisted that his forces and technologies were worse than those of the United States and Japan. At one point, he told the U.S. player when discussing carriers, "I'm the lightweight and you're the heavyweight." Nor did other countries, notably South Korea, give China the deference one would expect the dominant world power to command. Even the initial decision to play the wargame on a map of Northeast Asia suggested that the United States was still a global power and that China was still largely a regional power.

Second, the players did not necessarily adhere to the stated philosophy of employment, which was humans out of the loop. Before any moves were made, the players agreed that while the Chinese philosophy of employment might be humans out of the loop, the American philosophy would have humans in the loop. In practice, the China player rarely made decisions without humans in the loop, whereas the American players sometimes gave full tactical control to their machines.

Third, in the wargame, the commercial sector was supposed to be dominant relative to the government in AI. However, in practice, the players acted with little regard to the wishes of the player representing the commercial sector, and governments had unrestricted access to available AI technologies.

The Players Had Perfect Knowledge of Adversary Intent

Another wargame limitation arose from the nature of the game system. Although matrix gaming has significant advantages, such as simplicity, there turned out to be certain disadvantages for deterrence games. Using matrix gaming required free and open discussion among all players to determine the probability of success or failure for the actions undertaken. However, free and open discussion stifles key elements of deterrence wargames—namely, the chance for misperception and accident. Players were quite clear about what was being signaled, about what was a real show of force and what was not, and about the "offramps" that they built into their decisionmaking. Yet this is contrary to what Thomas Schelling explained as the defining characteristic of a game:[3]

> This is that at least two separate decision centers are involved, neither of which is privy to the other's planning and arguing, neither of which has complete access to the other's intelligence or background information, neither of which has any direct way of knowing everything that the other is deciding on. . . . What this mode of organization can do that cannot otherwise be done is to generate the phenomena of understanding and misunderstanding, perception and misperception, bargaining, demonstrations,

---

[3] Schelling was an early adopter of crisis gaming at RAND, a noted deterrence theorist, and a Nobel laureate in game theory.

dares and challenger's, accommodation, coercion and intimidation, conveyance of intent, and uncertainty about what each other has already done or decided on.[4]

In other words, the matrix game format itself may have reduced the misunderstanding and misperception possible in a wargame—and therefore the potential for inadvertent escalation. Only a few times did players gather for secret conferences, and, even when they did so, they subsequently explained their logic to the entire room. When North Korea fired its missile over Japan, the United States, Japan, and South Korea players knew that the missile was intended as a demonstration and so did not respond forcefully. Similarly, the North Korea player knew that the counterbattery fire that landed within his country was the decision of an AI and not sanctioned by national leaders. This remarkable event, a North Korean missile launch that was intercepted and a return missile launch that impacted within North Korean territory, petered out in the wargame because both sides knew the other's intentions perfectly.

Players Were Deliberately Aggressive

Because the wargame centered around deterrence and escalation, players were encouraged to be aggressive in ways that may be inconsistent with how they would act in real life. This introduced an artificiality into the game—one of many.

Players Had Control Over Autonomous Systems

One final limitation was that humans or random adjudication determined the behavior of the autonomous systems in the wargame. There are limits to how well humans can imitate machines, particularly in a case where misunderstanding between humans and machines could have important implications. At the same time, we acknowledge the immense technical challenge and resources required to replace every instance of human-played AI in the game with an actual one.

---

[4] Robert Levine, Thomas Schelling, and William Jones, *Crisis Games 27 Years Later: Plus C'est Déjà Vu*, Santa Monica, Calif.: RAND Corporation, P-7719, 1991, pp. 31–32.

CHAPTER SIX

Wargame Insights and Debates

What insights and observations did we gain from the wargame? What generated debates during the game between the players? We discuss them in this chapter.

Wargame Insights

Manned systems may be better for deterrence than unmanned ones. The presence of humans on some Chinese platforms made the U.S. and Japan players more hesitant to use force and often put the onus on them at several points to look for offramps in order to avoid further escalation.

Human casualties significantly increased tensions. Only unmanned systems were damaged or destroyed for much of the game. The first human deaths raised the stakes immediately. This raises the possibility of a future where ongoing, lower-level conflict between unmanned systems could persist for extended periods of time because the "cost" is lower than when human lives are at stake.[1]

Replacing manned systems with unmanned ones might not be seen as a reduced security commitment. Neither the Japan nor the South Korea players took the United States reducing manned forces in their countries as a sign of reduced security commitment. The South Korea player thought that neither country could keep up present-day levels

---

[1] One participant remarked that, although others had told him this was a possibility, he had not believed it until he saw it in the wargame.

of manning and even rated the ability of unmanned systems to monitor large areas as superior to humans. The U.S. and Japan players both regarded replacing U.S. Marines in Okinawa with robots as a positive development.

Settings on AI and autonomous systems can also be used to signal resolve and commitment during a conflict. Players in the game turned their defensive systems over to AI control to signal resolve. Japan, the United States, and South Korea all did this at points in the game in order to signal to China their willingness to use force. This was reminiscent of removing one's steering wheel and throwing it out the window in the game of "chicken" between two cars discussed in game theory.[2]

The speed of autonomous systems led to inadvertent escalation. In one incident, the U.S. and Japan players set their air defense systems to be fully autonomous and linked them to signal resolve to the China player. However, when North Korea unexpectedly launched a missile over Japan, the AI in the system not only shot down the missile but launched counterbattery fire that hit North Korea. Neither the United States nor Japan in the game intended to hit North Korea and potentially draw it into the conflict. That is, the players did not intend to use force, but the speed of machine decisionmaking led to the use of force.

Distances still mattered. Given the distances in the Pacific, players estimated that the events in the game would have played out over about a week. Distances alone, then, were still significant enough to slow down at least some of the potentially escalatory dynamics.

The presence of AI in decisionmaking created opportunities to confuse players and interject uncertainty into the dynamics. China's AI directed a single unmanned submarine to try and deter the use of a port. This appeared nonsensical to the U.S., Japan, and South Korea players. The U.S. players concluded that their side did not understand enough about China's AI to know what it was trying to do and stated that they would be taking time at that point to try and better understand it.

Players in the game had a mix of humans-in-the-loop, humans-on-the-loop, and humans-out-of-the-loop architectures. These architectures varied with the system and level (tactical versus operational) in ques-

---

[2] Schelling, 1966, p. 115.

tion. This is also likely to be reflected in real-world practice, where militaries may have a mix of system architectures and different levels of human oversight and control depending on the nature of the task.

U.S. players dramatically overestimated Chinese modernization. Players projected their force modernization, and the China player decided that China would have a mix of older, legacy systems and newer, modernized forces. However, the U.S. players planned for a highly modernized Chinese force and could not believe it when they encountered Chinese platforms that were immune to certain types of cyberattacks because they were simply too old. This raises an interesting question of what could happen if countries incorrectly predict adversary modernization in areas that are less observable before a conflict, such as AI.

Different valuation of human life and different philosophies of use led to specific escalatory dynamics. The United States in the game leaned heavily toward unmanned systems with a humans-in-the-loop philosophy that naturally slowed down the decisionmaking cycle. China chose to field more manned systems but had a philosophy of use under which machines made more (and faster) decisions, particularly at the tactical level. Players also assumed that the Chinese were more willing than the Americans to put their own people at risk. The U.S. players also stated that they would be less willing to fire on manned systems because of the lives at stake. This combination of factors led to several points in the game where the U.S. and Japan players attempted to diffuse a situation and explicitly look for offramps when in direct confrontation with the China player. With more humans in their decision cycle, and an aversion to targeting manned Chinese platforms, the U.S. and Japanese players ended up with the burden of trying to deescalate.

Debates During the Game

What Is the Singularity?

One debate during the game was over what the Chinese concept of the "singularity" should look like in practical terms. Some Chinese military theorists have proposed that the singularity is reached when the human brain can no longer cope with the ever-changing battlefield situation, and most of the time decision-making is given to highly intelligent machines. . . . In the end, human warriors will have to jump out of the chain of operations and smart machines will become the main force in future battlefields. The human war will form a new model of 'people on the loop.' . . . In the new model, people remain the ultimate decision makers.[3]

This concept stands in contrast to the American "Centaur" model, articulated by Bob Work, in which humans make decisions in conjunction with machines and are in the loop.[4]

For the wargame, we initially envisioned a strong, general intelligence AI that would control all of China's military systems in an enormous linked network. Some players proposed that such an AI was not enough—that a true singularity system would control all elements of national power, including trade policy and diplomatic tools. Others suggested that the original model went too far, and specifically that it was unrealistic to suggest that no humans would be involved in the decision to initiate conflict. Eventually, the team settled on Laoshi, a general intelligence AI that gathered information from all of China's linked military platforms and gave recommendations to human planners.

---

[3] Chen Hanghui, "Artificial Intelligence: Disruptively Changing the 'Rules of the Game,'" *China National Defense News*, March 18, 2016. The Chinese view of the singularity on the battlefield has also been characterized as the point "at which human cognition can no longer keep pace with the speed of decision-making and tempo of combat in future warfare" (Kania, 2017).

[4] Freedberg, 2015. The Chinese view of the singularity is also distinct from Vernor Vinge and Ray Kurzweil's singularity. Vinge described the singularity as the point where greater-than-human intelligence allows for unprecedented developments (Vernor Vinge, "The Coming Technological Singularity: How to Survive in the Post-Human Era," paper presented at VISION-21 Symposium sponsored by the NASA Lewis Research Center and the Ohio Aerospace Institute, March 30–31, 1993). Kurzweil's singularity is a "future period during which the pace of technological change will be so rapid, its impact so deep, that human life will be irreversibly transformed" (Ray Kurzweil, *The Singularity Is Near: When Humans Transcend Biology*, New York: Penguin Books, 2005).

Under What Conditions Could Humans Be Taken Out of the Loop?

At the beginning of the wargame, there was general consensus that two factors determined when humans will be taken out of the loop: (1) the vulnerability of the military asset involved and (2) the potential consequences of an accident. In situations where expensive assets were at risk of imminent destruction and in situations where an autonomous system activating in error would have minimal negative consequences, players were more willing to go "full auto." As an example, players often and easily made the decision to turn over all control of antimissile defenses to the thinking machines.

As the wargame progressed, players discussed other situations where humans might take themselves out of the loop. One player argued that a nation facing serious consequences from losing a conflict might be willing to give complete control of its military assets to thinking machines in a desperate attempt to gain an advantage. In such a case, a country's philosophy of use concerning AI might change at a certain point in a conflict, but in ways not immediately obvious to its adversary.

Players also argued that autonomous systems might become so commonplace that humans trust them to operate without interference. In this potential world, to the extent that humans remain in the loop, they act only to run diagnostics to make sure that the software is working as specified. As one player stated, humans might "stop asking why the system is working and just worry about whether the machinery is working."

As an analogy, one player used the probable path of the relationship between humans and self-driving cars. Today, he posited, if someone came into work and said, "I got in an accident last night—I was letting my car drive itself," the response from his coworkers would be, "What were you thinking? Why was the machine driving?" However, the player argued that in ten years, if someone came into work and said, "I got in an accident last night—I decided to do the driving myself," the response would be, "What were you thinking? Why was the machine not driving?" As AI demonstrates its capabilities, humans will grow to trust and possibly overtrust machines. War, the player argued, will be no different.

Other players expressed doubt at this proposition, with one saying that there is a difference between self-driving cars and war. If your car decided to drive into some children, he questioned, would you still trust it? Other players responded that people already trust machines despite errors—there have been numerous instances of people following the directions from GPS software right into a lake. One player confessed that he had followed his GPS into a restricted military base.

What Is the Probability of an Accident? How Will Autonomous Systems Behave?

The debate over the likelihood of unintentional, autonomous fires became a debate about the probable behavior of autonomous systems in general. In the wargame, most instances in which a weapon system went "full auto" occurred without incident. We assessed that for routine events—such as the escorting of adversary aircraft—the chance of accident would be low. However, during a more complicated event, the U.S. antimissile system automatically launched counterbattery fire into North Korea in response to a demonstration, something the human players did not order.[5]

Determining the likelihood of accident rested upon several questions. First, how would humans program AI? Even in a world where autonomous systems learn and perform actions that they were not explicitly programmed to do, humans still maintain a great deal of control, particularly in setting the goal of an AI. If humans are most concerned with avoiding accidental conflict and program AI to be correspondingly cautious, accidents caused by autonomous systems could be less likely. On the other hand, if humans value the quickest response times possible and program AI to be aggressive, then accidents such as unintentional fires could be much more likely.

Second, what would the balance between sensing and concealing technologies look like in the future? Players largely agreed that both sensor and stealth technologies will likely improve in the coming

---

[5] To clarify how this occurred, when the North Koreans launched their missile, a player speculated that the automated defense grid might launch counterbattery fire. After debate, a probability was assigned to this event, and a roll of the dice indicated that it happened.

decades, but the team felt that it could not give a reasonable assessment as to which would dominate.

Third, what would AI learn about human conflict? Some players argued that crises between nations are so rare and so different that a learning program could not develop generalizable strategies from their study. Others wondered whether AI might develop aggressive strategies from a study of conflict, which would create a technological "cult of the offensive."[6]

Fourth, how might AI behave in ways which are totally unpredictable to humans? AI has reached the stage of technological maturity in certain domains where it can arrive at outcomes not based upon human experience. An example brought up during the wargame was Google's new game program, AlphaZero. Unlike other chess programs, AlphaZero never analyzed human games and instead taught itself how to play chess by playing against itself repeatedly. The strategies it developed were unlike anything ever seen before on a chessboard— completely baffling to humans and traditional computer programs alike. One chess grandmaster stated, "I always wondered how it would be if a superior species landed on earth and showed us how they play chess. I feel now I know."[7]

Will the Lower Cost of Unmanned Systems Increase Conflict?

The team generally agreed that unmanned systems provide opportunities for decisionmakers to accept greater risks because there could be lower consequences to a drone being destroyed than a pilot being shot down. Players disagreed, however, about what those lower consequences would mean for deterrence.

One perspective was that the lower costs of losing unmanned systems made war less likely because national leaders would be less motivated to save face when systems are destroyed. These players pointed to

---

[6] Stephen Van Evera, "The Cult of the Offensive and the Origins of the First World War," in Steven E. Miller, Sean M. Lynn-Jones, and Stephen Van Evera, eds., *Military Strategy and the Origins of the First World War*, Princeton, N.J.: Princeton University Press, 1985.

[7] Mike Klein, "Google's AlphaZero Destroys Stockfish in 100-Game Match," Chess.com, December 6, 2017.

the differing escalation dynamics that occurred within the game. After an unmanned Japanese cargo ship was destroyed, the Japan player did not feel that he had to take kinetic actions—the player felt that that he still had other options available. However, after the Chinese lost a manned submarine, the China player felt that he had to escalate in order to maintain his regime's popularity with the Chinese people.

A second view was that the lower costs of losing unmanned systems would lead to chronic low-level conflict. Just as certain military actions have become routine, such as the testing of an adversary's airspace with fighters and the subsequent intercept and escort mission, so too would the daily destruction of a few unmanned systems. One player even noted, "You can imagine in America that an unmanned system being shot down would be less important than people's daily Starbucks."

A third perspective took a different lesson from the example of the Chinese submarine. In this view, the reduced cost of losing unmanned systems meant that nations would strategically keep certain platforms manned in order to improve deterrent credibility, even if the manned platforms were less effective than fully autonomous systems. One player, as an example, recounted the history of American ISR efforts over Iran. Iranian leaders threatened to shoot down American drones. In response, the United States flew manned escort missions alongside the unmanned drones in order to increase the costs of Iranian aggression. In this real-world example, a country deliberately put human life at risk to prevent the destruction of unmanned assets. Another player suggested that unmanned and manned systems would be developed in concert so that adversaries would have no way of knowing whether a particular platform was manned or unmanned, so as to increase its deterrent value compared with that of a clearly unmanned one.[8]

---

[8] A 2018 RAND simulation effort for the Army Science Board Panel on Manned-Unmanned Teaming examined the potential effectiveness and burden of some of these strategies. Randall Steeb and Morgan Kisselburg, "Counter-IADS MUM-T Exploratory Quantitative Analysis," RAND Briefing to 2018 Army Science Board Panel on MUM-T, July 20, 2018.

CHAPTER SEVEN

Implications for Deterrence

What are some further implications for deterrence from the greater proliferation of autonomous and unmanned systems? After running our wargame as a thought experiment, what might we be able to say more broadly about the potential implications? In this chapter, we revisit certain deterrence concepts discussed in Chapter Two, make comparisons with nuclear deterrence, discuss the role of presence, and lay out some initial ideas about signaling and understanding.

Deterrence Concepts Revisited

We saw in our game that autonomous and unmanned systems have the potential to affect **extended deterrence** and our ability to **assure** our allies of U.S. commitment to their defense. In this specific wargame, the players representing U.S. allies did not perceive a lesser U.S. security commitment when the U.S. player changed out U.S. troops in Japan and South Korea for robots. However, it will be important to ask whether this is will be true in other cases.

On the one hand, autonomous systems could enhance the credibility of U.S. conventional extended deterrence because the risk to U.S. military personnel of employing these systems is much lower than with traditional kinetic military means. Additionally, the operational advantages of autonomous systems relative to more traditional military means—such as faster decision cycles, the ability to stay ready to strike much longer, and greater precision than human personnel—could lead adversaries and allies to conclude that U.S. leaders will be more willing and likely to employ autonomous systems in situations in which allied interests are threatened.

On the other hand, U.S. allies could interpret Washington's increased reliance on autonomous systems as a reflection of a growing U.S. unwillingness to put American lives on the line in severe crises and confrontations with adversaries. Thus, although the United States may see fielding autonomous systems as a way to reduce risk to U.S. military personnel by substituting machines for humans, reducing risk to U.S. personnel in overseas commitments may paradoxically reduce Washington's ability to assure U.S. allies.

Autonomous systems may also affect the **credibility** of deterrent threats.[1] States with autonomous systems might appear more credible when making deterrent threats than states without them.[2] Nonetheless, as with other conventional weapons, opponents who do not possess autonomous systems will not simply accede to the deterrent or coercive threats of states that do have them. Instead, they will develop strategies, operational approaches, and capabilities designed to counter, avoid, or mitigate the advantages of autonomous systems. When confronting states that do possess autonomous systems of their own, using autonomous systems could come to be seen as low-risk and thus attractive means for mounting probing attacks against adversaries. This could result in "salami" tactics employed to slice away at the adversary's interests without overtly crossing a threshold or red line that invites the opponent to strike back.

Widespread AI and autonomous systems could also make **escalation** and **crisis instability** more likely by creating dynamics conducive to rapid and unintended escalation of crises and conflicts. This is because of how quickly decisions may be made and actions taken if more is being done at machine, rather than human, speeds. **Inadvertent escalation** could be a real concern.

In protracted crises and conflicts between major states, such as the United States and China, there may be strong incentives for each

---

[1] It is also important to draw a distinction between credibility of **capability** and credibility of **will** (or resolve).

[2] See Boyle, 2014, p. 78, for a discussion of the impact of the arms race in military drones.

side to use such autonomous capabilities early and extensively, both to gain coercive and military advantage and to attempt to prevent the other side from gaining advantage.[3] This would raise the possibility of **first-strike instability.**

AI and autonomous systems may also reduce **strategic stability.** Since 2014, the strategic relationships between the United States and Russia and between the United States and China have each grown far more strained. Countries are attempting to leverage AI and develop autonomous systems against this strategic context of strained relations. By lowering the costs or risks of using lethal force, autonomous systems could make the use of force easier and more likely and armed conflict more frequent.[4] A case may be made that AI and autonomous systems are destabilizing because they are both transformative and disruptive. We can already see that systems such as UAVs, smart munitions, and loitering weapons have the potential to alter the speed, reach, endurance, cost, tactics, and burdens of fielded units.

Additionally, AI and autonomous systems could lead to **arms race instability.** An arms race in autonomous systems between the United States and China appears imminent and will likely bring with it the instability associated with arms races. Finally, in a textbook case of the **security dilemma,** the proliferation of autonomous systems could ignite a serious search for countermeasures that exacerbate uncertainties and concerns that leave countries feeling less secure.

Comparisons with Nuclear Weapon Proliferation

In this section, we briefly revisit ideas about nuclear weapon proliferation in comparison to autonomous systems. Using the same concepts for autonomous systems as for nuclear weapons is not correct, just as

---

[3] James N. Miller, Jr., and Richard Fontaine, *A New Era in U.S.-Russian Strategic Stability*, Cambridge, Mass.: Belfer Center, Harvard Kennedy School and Center for New American Security, September 2017.

[4] For a summary of similar arguments made about armed drones, see Michael C. Horowitz, Sarah E. Kreps, and Matthew Fuhrmann, "Separating Fact from Fiction in the Debate over Drone Proliferation," *International Security*, Vol. 41, No. 2, Fall 2016, pp. 13–14.

understanding machine guns within the paradigm of cannons is not correct. New technology often calls for new paradigms. However, until we can identify these better paradigms, it can sometimes be helpful to use existing ones as a point of departure. Comparing nuclear weapon proliferation with the spread of autonomous systems, we postulate in Table 7.1 their differences along a number of dimensions. Overall, we expect much wider proliferation and use of autonomous weapon systems.

Several factors discouraged the use and proliferation of nuclear weapons: very high barriers to entry, slow proliferation, reluctance to use nuclear weapons, low incentives for first use once second-strike capabilities became available, and well-developed verification regimes.

These factors do not apply to autonomous systems. There are significantly lower barriers to the development and the use of autonomous systems. In addition, verification regimes to prevent autonomous systems proliferation are not feasible. We expect actors to be willing to use autonomous systems first if they perceive that doing so offers

Table 7.1
Nuclear Weapon Proliferation Versus Autonomous System Proliferation

Dimension	Nuclear Weapons	Autonomous Systems
Development and use	State-led/single application	Commercial-led/dual-use
Barriers to entry	High and punitive	Modest
Horizontal proliferation	Slower than predicted	Very fast
Vertical proliferation	Arms racing has happened between the United States and Soviet Union	Arms races in military autonomous systems are shaping up across states and nonstate actors
Willingness to use	Low/only in extreme situations	Very high
Incentives for first use	Historically low once viable second-strike capabilities are in place	Likely very high
Active defenses against	Generally, still infeasible	Likely feasible
Verification regimes	Well developed	Likely infeasible
Command and control	Highly centralized	Highly decentralized

an advantage—as has been the case with remotely piloted unmanned aerial systems. Another difference is that the command and control of nuclear weapons, with the exception of ballistic missile submarines, is highly centralized and has many checks. With autonomous systems, we expect command and control to be the opposite: highly decentralized, with far fewer checks on use.

How Escalatory Dynamics May Change

In this section, we explore some more general ideas prompted by the wargame that also have the potential to affect deterrent and escalatory dynamics. We hypothesize that the different mixes of humans and artificial agents in different roles can affect the escalatory dynamics between two sides in a crisis. We also examine how signaling and understanding, important elements to successful deterrence, could be adversely affected with the introduction of machine decisionmaking.

Decisionmaking and Presence

One insight from our wargame is that the differences in the ways two sides configure their human versus machine decisionmaking and their manned versus unmanned presence could affect escalatory dynamics during a crisis. In the wargame, confrontations occurred between unmanned U.S. forces with humans-in-the-loop decisionmaking and Chinese forces that were manned but had more humans-on-the-loop and humans-out-of-the-loop decisionmaking. These confrontations appeared to put the onus on U.S. forces to deescalate the situation and inspired Table 7.2. We hypothesize the further ways that mixes of human and machine could result in different escalatory dynamics.

In the upper left of Table 7.2, we propose that when systems are manned and the decisionmaking is primarily done by humans, there is a lower escalatory dynamic. We argue that humans in the decisionmaking process have time to slow down how quickly things can escalate, but that the presence of humans means that there is a higher cost to miscalculating events, because human lives could be lost. This quadrant represents the most common situation today.

Table 7.2
Human and Machine Configurations and Potential Escalatory Dynamics

| | | Decisionmaking | |
		Primarily Human	Primarily Machine
Physical Presence	Human	Lower escalatory dynamic Higher cost of miscalculation	Higher escalatory dynamic Higher cost of miscalculation
	Machine	Lower escalatory dynamic Lower cost of miscalculation	Higher escalatory dynamic Lower cost of miscalculation

In the lower left-hand quadrant, there are primarily unmanned systems with primarily human decisionmaking—this may be the least escalatory combination of all. Having humans in the loop again slows down the decision cycle compared with configurations that are more heavily driven by machine decisions, which may mean more time to consider deescalatory offramps during a crisis. Humans may also be better at understanding signaling. Additionally, having mostly unmanned systems lowers the risk to human life, as the consequences of miscalculating are destroyed systems but not loss of human lives. This is the quadrant that best represents the United States in the wargame in Chapter Five.

In the upper right is a situation in which systems are manned but decisions are made mostly by machines (humans on the loop or humans out of the loop). We argue that this is the most escalatory situation of all. With more decisions happening at machine speeds, there is likely a greater risk of inadvertent escalation during a crisis. However, the presence of humans means that there is the higher risk to human life with miscalculation and escalation. This is where the notional, future Chinese forces were in the wargame.

In the final, lower right-hand quadrant, we see the combination of unmanned systems and machine decisionmaking. This is perhaps what the public imagines futuristic war will be like one day. We argue that this has a higher escalatory dynamic because of the rapid machine decisionmaking, but the costs of miscalculation are lower because human lives are not at risk.

Escalatory dynamics might change considerably when adversaries from different quadrants come into conflict. In the wargame, the United States was operating from the lower left quadrant and China from the upper right. In the game, the result of this interchange was several attempts by the United States and Japan to deescalate the confrontation. It may be that this particular combination may give rise to a world where the onus is on the United States to constantly try to deescalate because (1) the United States is unwilling to kill humans because of its perceived escalatory effects and (2) greater human decisionmaking can include considerations about escalation in ways that may elude machines. The implication, however, is that hypothetical adversary forces using machine decisionmaking with humans on board may do better in a game of chicken against these hypothetical U.S. forces, meaning that U.S. forces would need to keep backing down or look for ways to diffuse the situation. This is an important consideration as the United States continues the path of humans-in-the-loop unmanned systems.

We encourage the reader to consider the different potential dynamics of a two-sided conflict with sides representing different quadrants in Table 7.2.

Signaling and Nonhuman Decisionmaking

What happens to signaling when not only humans but also machines are involved in sending and receiving signals? In a world with only humans doing the signaling, they do try to show resolve and communicate a deterrent threat, but they also seek to avoid further escalation and to deescalate conflicts. How will machines interpret such signals? Theory of mind, demonstrated from an early age, allows humans to understand that other humans may hold intentions and beliefs about a situation that are different from what they themselves hold to be true. It is this natural ability in most humans that allows them to make some predictions about the behavior of others.[5] There is the chance that statistical machine learning could predict certain behaviors from signals,

---

[5] Brittany N. Thompson, "Theory of Mind: Understanding Others in a Social World," *Psychology Today*, blog post, July 3, 2017.

but a very important question in this regard is what data have been used to train the models that make it to the battlefield.

We acknowledge the numerous historical cases in which humans have misinterpreted signals from other humans. However, we argue that machines, on the whole, are still worse at understanding intended human signals than are humans, particularly because there is often a complex context that the machine will not understand. We also argue that machines lack theory of mind in novel situations with humans.

Table 7.3 lists some of our hypotheses about how machines that are programmed to take advantage of changes in the tactical and operational picture might react to different human signals. Key here is the idea that machines that are set up to rapidly act on advantages they see developing on the battlefield may miss deescalatory signals. In other words, signals developed over decades between humans to deter or deescalate a conflict could have the opposite effect and rapidly escalate a situation if machines are not programmed, or taught, to take deterrence and deescalation into consideration. AI that is set to be aggressive may be at greater danger of misreading the intent behind such signals.

We see in Table 7.3 that autonomous systems, programmed to take advantage of tactical and operational advantages as soon as they can identify them, might create inadvertent escalation in situations where the adversary could be trying to prevent further conflict and escalation. We are not arguing against implementing systems that

Table 7.3
Potential Machine Misinterpretation of Human Signaling

Human Signal	Signal's Purpose	Potential Machine Interpretations	Potential Machine Actions
Increase alert status or presence of forces	Indicate resolve and deter attack	• Deteriorating tactical or operational situation • Signs of imminent attack	• Increase alert status or presence of own forces • Preemptive strike
Continue present actions	Indicate intent not to further escalate	• Adversary failing to take key defensive actions	• Position forces to take advantage of opportunities
Reduce alert level or presences of forces	Indicate a desire to deescalate a situation	• Improving tactical or operational situation • Key opportunities are now present	• Position forces to take advantage of opportunities • Strike adversary

can quickly identify opportunities on the battlefield. It is, however, advisable to ask how to review the situation for adversary signals that machines may miss.

Level of Understanding

Understanding of an adversary's will, resolve, and intent are central to deterrence. Figure 7.1 is a simplified diagram of how deterrence has traditionally worked: humans signaling to, interpreting, understanding, and anticipating other humans. (We use blue to denote friendly forces and red to denote adversarial ones.) Put in simple terms, traditional deterrence primarily required humans understanding other humans.

In Figure 7.2, we add the types of understanding that are required once machines are involved. Not only must humans understand adversary humans as in Figure 7.1, the following must also occur:

- Humans understand their own machines.
- Humans understand adversary machines.
- Machines understand their humans.
- Machines understand adversary humans.
- Machines understand other machines.

Misunderstanding along any of these dimensions introduces possibilities for misinterpretation, misperception, and miscalculation.

Figure 7.1
Understanding Required in Traditional Deterrence

Figure 7.2
Understanding Required in Deterrence with AI and Autonomous Systems

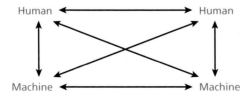

Humans understanding their own machines and their range of potential behaviors is not a trivial undertaking. We already have historical examples of systems such as the Phalanx antimissile system firing on U.S. ships and aircraft in ways not anticipated by their human operators and killing U.S. servicemen.[6]

An even more difficult problem is humans trying to understand adversary machines, particularly machine learning systems. The first obvious problem is that humans do not have ready access to adversary algorithms or understanding of how an adversary system is programmed. For learning systems, even if the algorithm is known, it may be impossible to know the data on which the system trained. Even if the algorithms and data are somehow known, how the adversary intends to use the system and under what circumstances may still be unknown. That is, adversary human-machine collaboration may be a mystery. Rather, humans on one side of the equation may be left trying to infer intent and potential behavior from partial observations.

Understanding between humans and machines is a two-way street. It is necessary for machines to accurately understand the intent of their own humans, adversary human behavior and intent, and adversary machine behavior in order to avoid misunderstanding and miscalculation. Will machines accurately understand adversary humans and machines if the adversary behaves differently during conflict, when most of the data on the adversary were collected during peacetime? If machines understand the future primarily through correlation, will they appropriately correlate unexpected adversary behaviors to the "right" things?

Figure 7.2 becomes even more complicated when allied and coalition partners and their machines enter the picture. Interoperability with learning systems will pose challenges. And this does not even begin to address a future with a large number of autonomous civilian machines also operating throughout the environment.

---

[6] Paul J. Springer, *Outsourcing War to Machines: The Military Robotics Revolution*, Praeger Security International, 2018.

CHAPTER EIGHT

Implications for Decisionmaking

Underlying many of the ideas in the previous chapter are the nuts-and-bolts mechanics of human and machine interactions. How could these interactions set the stage for accidents or escalation? We discuss four main topics in this chapter that relate to these questions: inadvertent engagement, trust in autonomy, the effects of time scale on decisionmaking, and considerations to keep in mind when building AI and autonomous systems more broadly into military forces.

Inadvertent Engagement

Our wargame demonstrated an instance of inadvertent engagement on the part of autonomous systems. This is a topic that warrants further consideration, because of the escalatory potential from unintended engagement or use of force. There are numerous human examples of friendly fire and inadvertent civilian deaths that have not involved autonomous systems. However, it is worth examining how autonomous systems might exacerbate the problem of inadvertent engagement. In this section, we provide a very brief history of inadvertent engagement with systems that already exist today with less direct human involvement, then we discuss the possible implications of even more advanced, autonomous systems.

Past Inadvertent Engagements by Autonomous Systems
Military autonomous systems are not new, and neither is inadvertent engagement by such systems. Examples include landmines, torpedoes,

close-in weapon systems such as Phalanx,[1] and area defense systems such as Aegis. In use since the U.S. Civil War,[2] landmines are unable to distinguish among friendly forces, adversary forces, and civilians.[3] At least two German U-boats are believed to have been sunk by their own acoustically homing torpedoes during World War II.[4] There were also many cases of "circular runs" by American torpedoes in World War II in which torpedoes circled back toward the submarines that launched them. The USS *Tang* and the USS *Tullibee* were sunk by their own torpedoes.[5] The threat of a circular run by a torpedo persists today; it is mitigated by procedures and the capability to guide torpedoes after launch.[6]

Phalanx has also experienced several mishaps. In 1989, the USS *El Paso* used Phalanx to destroy a target drone. The drone fell into the sea, but the Phalanx reengaged it as it fell and struck the bridge of the nearby USS *Iwo Jima,* killing one and injuring another. During the 1991 Gulf War, the USS *Missouri* launched chaff to confuse an incoming Iraqi missile. The Phalanx system on the nearby USS *Jarrett* shot at the *Missouri*'s chaff and hit the ship four times.[7] In 1996, a U.S. A-6E Intruder aircraft towing a radar target during gunnery exercises was shot down when a Phalanx aboard the Japanese destroyer *Yūgiri* locked

---

[1] Raytheon, "Phalanx Close-In Weapon System: Last Line of Defense for Air, Land and Sea," webpage, undated.

[2] John Grady, "Mine Warfare in the Civil War," Army History Center, December 9, 2016.

[3] Geneva International Centre for Humanitarian Demining and Stockholm International Peace Research Institute, *Global Mapping and Analysis of Anti-Vehicle Mine Incidents in 2017,* April 2018; "Landmines Killed More Than 2,000 People in 2016," *Deutsche Welle,* December 14, 2017.

[4] Guðmundur Helgason, "The Torpedoes," uboat.net, 2018.

[5] Dejan Milivojevic, "US WW2 Sub Sunk Itself When Its Own Torpedo Made a Full Circle & Struck It," *War History Online,* January 30, 2019.

[6] "MK 48 Mod 7 Common Broadband Advanced Sonar System (CBASS) Heavyweight Torpedo," *Naval Technology,* undated.

[7] Springer, 2018.

onto the A-6E instead of the target. A post-accident investigation concluded that the *Yūgiri*'s gunnery officer gave the order to fire too early.[8]

Aegis has been involved in an especially high-profile case of inadvertent engagement. In 1988, the Aegis cruiser USS *Vincennes* mistook an Iranian civilian airliner for an Iranian fighter and shot it down. It fired two surface-to-air missiles at the airliner and killed all 290 crew and passengers aboard.[9] There had been hostilities prior to the incident between U.S. and Iranian forces, including the USS *Samuel B. Roberts* striking a mine and Iranian forces firing on U.S. helicopters. New U.S. rules of engagement also authorized positive protection measures before coming under fire.[10] After *Vincennes* inadvertently crossed into Iranian waters, Revolutionary Guard gunboats fired on *Vincennes*'s helicopter. The *Vincennes* crew also erroneously concluded that the airliner was descending toward the *Vincennes* when it was in fact climbing.[11] These and other factors led to the *Vincennes* firing on the airliner. A review of the incident by the Chairman of the Joint Chiefs of Staff concluded that while errors had been made, the captain and crew had acted reasonably. The review also found that the Aegis system had performed as designed—particularly, that it was "never advertised as being capable of identifying the type of aircraft being tracked. That decision is still a matter for human judgment." However, one recommendation was to improve the Aegis display systems in order to better identify important data.[12]

Table 8.1 summarizes these mistaken engagements with autonomous systems. We note the type of system, the nature of the incident, and reasons behind the mishap. Common reasons for mistaken

---

[8] Philip Shenon, "Japanese Down Navy Plane in an Accident; Crew Is Safe," *New York Times*, June 5, 1996. Both crew members ejected safely.

[9] Max Fisher, "The Forgotten Story of Iran Air Flight 655," *Washington Post*, October 16, 2013.

[10] William M. Fogarty, *Formal Investigation into the Circumstances Surrounding the Downing of Iran Air Flight 655 on July 3, 1988*, DTIC AD-A203 577, July 28, 1988, pp. 1–8.

[11] Fogarty, 1988, p. 5.

[12] Fogarty, 1988, pp. 6–8.

Table 8.1
Inadvertent Engagements by Autonomous Systems

System	Time Frame	Circumstances	Reasons for Mishap
Landmines	Continual	Postconflict civilian casualties	• Persistence after a conflict • System unable to distinguish between combatants and noncombatants
Torpedoes	World War II	German and U.S. submarines sunk by their own torpedoes	• System unable to distinguish between friendly and adversary forces
	World War II	"Circular runs" by U.S. torpedoes	• System unable to distinguish between friendly and adversary forces
Aegis	1988	USS *Vincennes* shoots down Iranian civilian airliner	• Humans misidentify target • System unable to distinguish between types of aircraft • Possible human confusion with Aegis data display
Phalanx	1989	Phalanx reengages after destroying target drone and hits friendly ship	• System unable to account for nearby friendly forces as it fires at target
	1991	Phalanx fires at friendly chaff and strikes friendly ship	• System misidentifies target • System unable to account for nearby friendly forces as it fires at target
	1996	Phalanx shoots down friendly aircraft towing exercise target	• Human gives order to fire too soon • System misidentifies target

engagements include target misidentification, an inability on the part of the system to account for friendly forces, and human error.

Implications of More-Advanced Autonomous Systems

We expect future autonomous systems to be more capable in a number of ways. This could include increased pattern recognition from statistical machine learning to improve target recognition and reduce risks during target selection. It could also involve improved sensing to shorten the decision cycles by which autonomous systems move from searching for and acquiring targets, to engaging them, to deciding to disengage. The proliferation of these more capable systems will likely increase the frequency of their use.

How could inadvertent engagements such as those we discussed in Table 8.1 change with more-widespread and more-advanced systems?[13] On the one hand, better AI could reduce mistaken engagements through improved target identification, addressing the current problem of discriminating between targets and nontargets. On the other hand, we have noted cases where human error contributed to mishaps. Human error interacting with even more-complex systems could very well contribute to future mistaken engagements.[14] Lastly, Table 8.1 largely covers autonomous systems in naval environments with limited civilian presence. Even as AI could improve differentiating targets from nontargets, having more autonomous systems on the ground and in populated areas may come with significant challenges in accounting for friendly forces and noncombatants. We present some potential advantages and disadvantages of future systems in Table 8.2.

What are the implications of more-advanced and more-widespread autonomous systems for deterrence and escalation? As more-complex systems and more-complicated human-machine interactions develop,

Table 8.2
Potential Advantages and Disadvantages of Military Autonomous Systems

Generation	Advantages	Disadvantages
Current systems	• Persistence • Force multiplier	• Difficulty discriminating among targets • Still affected by human error
Future systems	• Potential improvements in target recognition • Force multiplier	• Increasing system and environment complexity could give rise to new types of errors • Learning systems will exhibit behavior not necessarily seen in testing • Continued possibilities for human error

[13] For additional thoughts on the issue, see Paul Scharre, *Autonomous Weapons and Operational Risk*, Washington, D.C.: Center for New American Security, 2016, pp. 18–22.

[14] Freedberg, 2018a.

there is clearly the possibility of technical accidents and failures. There is the possibility that one side may interpret accidental engagements by autonomous systems as deliberately escalatory or even preemptive in nature. This is particularly true because it is extremely difficult to surface the full range of behaviors that autonomous systems are capable of during testing. On the other hand, timely notification about accidents and inadvertent engagements, perhaps communicated through means or channels worked out in advance, could help avoid misinterpretation and escalation.

Trust and Autonomy

An important topic in robotics research is trust in autonomy.[15] Although this topic has often come up in relation to recent developments in self-driving vehicles, it dates back to the earliest days of automated systems. The notions of "overtrust" and "distrust" and the need for "trust calibration" in human-machine collaboration are critical to understanding whether users will be able to deploy and use systems effectively. These concepts are also critical to determining whether autonomous systems can be effectively tested, verified, and validated.[16]

Trust also fundamentally affects not just whether and how systems will be employed, but also how they could influence decisionmaking. Although it will depend on the details of the system and the nature of the human-machine collaboration in decisionmaking, we argue that overtrust or mistrust in autonomous systems can increase escalatory uncertainty and result in missed opportunities for deescalation. This is because overtrust could lead to reduced human monitoring of or intervention in complex situations or machine target selection. The overall result might be a greater likelihood of engaging in lethal actions with-

---

[15] John D. Lee and Katrina A. See, "Trust in Automation: Designing for Appropriate Reliance," *Human Factors*, Vol. 46, No. 1, March 2004, pp. 50–80; Munjal Desai, *Modeling Trust to Improve Human-Robot Interaction*, dissertation, University of Massachusetts Lowell, 2012; Mittu Ranjeev, Donald Sofge, Alan Wagner, and W. F. Lawless, *Robust Intelligence and Trust in Autonomous Systems*, Boston, Mass.: Springer, 2016.

[16] Ranjeev et al., 2016.

out the full benefit of the human oversight that was intended. As for misplaced mistrust in systems, while this may impair some of the benefit of having systems that are able to respond faster than humans, we hypothesize that additional human scrutiny of machine decisionmaking and activity should lengthen response times and provide additional opportunities to deescalate a quickly evolving crisis.

Time Scale and Decisionmaking

Issues of trust outlined above notwithstanding, the reduced decisionmaking and reaction times that come with AI and autonomous systems are considered a key operational advantage when it comes to improving the lethality of a system or an effects chain. The trade-off between these two objectives—maintaining sufficient awareness and control of AI and autonomous systems to manage escalation, while also taking advantage of AI to improve the lethality of one's military capabilities—will be a central concern as AI and autonomy proliferate. In this section, we discuss different ways to consider how machine decisionmaking and time affect these trade-offs.

A Decision Cycle—Observe, Orient, Decide, and Act

It may be helpful to return to a familiar model of the decision cycle in the military: Boyd's observe, orient, decide, and act (OODA) loop, depicted in Figure 8.1.[17] At the heart of Boyd's concept is the idea of simultaneously compressing one's own time and stretching out an adversary's time to generate favorable mismatches in time and space. Doing this with confidence requires understanding how the adversary observes, orients, and decides to act.

Boyd's OODA model is one way for researchers and planners to speed up their own decision cycle by considering at what stage in the cycle AI or autonomous systems may shorten the processes. It may also be used as a framework to evaluate where adversary AI and autonomous systems could greatly speed up their decision cycles.

---

[17] Boyd, 2018; Osinga, 2005.

Figure 8.1
Boyd's Observe, Orient, Decide, Act (OODA) Loop

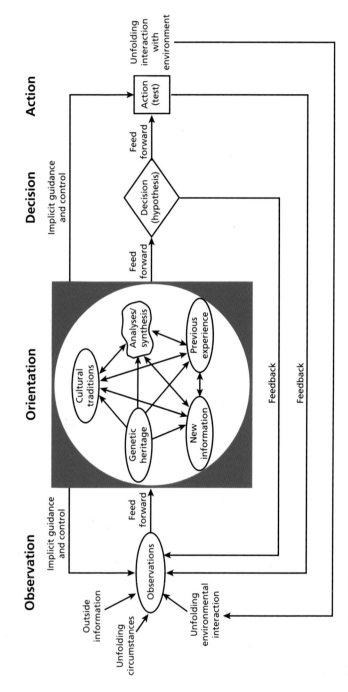

SOURCE: Adapted from Boyd, 2018.

During our workshop, it became evident that one would need to consider OODA cycles at the strategic, operational, *and* tactical levels. This is because how AI may be employed and integrated within decision cycles, and what types of sensing and information sources feed an AI's machine learning processes, will likely differ by level. For example, consider the possibility of an adversary using humans-on-the-loop AI in its operational planning processes, but humans-out-of-the-loop AI in certain tactical mission areas.

Decisions by Time Scale

Another way to think about time is to consider reaction and type of autonomy in the system. Figure 8.2 maps out some weapon capabilities and roughly where they fall in a continuum of reaction speeds versus type of autonomy. Reaction speeds range from machine speeds that are fractions of a second to longer time lines that could stretch into weeks and years. The autonomy axis goes from autonomy at rest to autonomy in motion, although systems can certainly be combinations. The purpose of Figure 8.2 is to help map out where a specific autonomous system may be along both axes, in order to characterize and compare reaction times and whether the system's effects manifest themselves primarily in the physical or virtual worlds. Inadvertent escalation may be a greater danger from systems that have extremely short reaction times, while autonomous systems that are primarily physical may change the cost calculus for countries that employ them.

Centralized-Decentralized Command and Control Versus Human-Machine Interaction

Yet another way to organize time, and potential decisionmaking on the battlefield and its implications, is to differentiate between centralized or decentralized decisionmaking and human or machine decisionmaking. We often think of command and control as being either centralized or decentralized among human units, with decentralized decisionmaking sometimes referred to as "mission type" orders.[18] Yet data and

---

[18] Robert W. Peterman, *Mission-Type Orders: An Employment Concept for the Future*, Maxwell Air Force Base, Ala.: Air University Press, 1990, pp. 3–5.

Figure 8.2
Reaction Speeds and Type of Autonomy

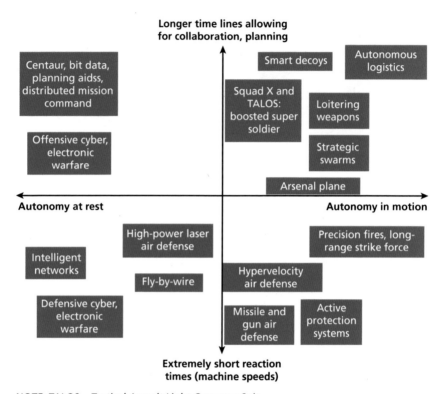

NOTE: TALOS = Tactical Assault Light Operator Suit.

computing can also be centralized or decentralized. Edge computing is processing data as close to the source as possible, rather than bringing data back to a more central location for processing, and is a current trend in commercial computing.[19] This is because the proliferation of commercial sensors and data through online sources can lead to greater latency between bringing data back to a local or cloud platform, ana-

---

[19] Jon Markman, "This Is Why You Need to Learn About Edge Computing," *Forbes*, April 3, 2018.

lyzing it, and taking action.[20] Bandwidth limitations on bringing back increasing amounts of data to be centrally processed are also a real issue.[21]

Edge computing architectures for autonomous systems means that they will make decisions with what they can sense or process locally—and with less than the full amount of information available. Likewise, it will be impossible for central AI on the battlefield to have real-time access to the data that all sensors are collecting. Therefore, in order to act on a developing situation in as timely a manner as possible, autonomous systems may likely be acting on their local understanding of what is going on. Truly centralized AI to support decisionmaking may be much slower than what the tempo of the battlefield allows.

Time, Deterrence, and Escalation

How might decision cycle times and centralized or decentralized machine decisionmaking affect deterrence and escalation? Overall, significantly shorter decision cycles of the type envisioned for future AI-enabled militaries could significantly speed up inadvertent escalation once some course of action is triggered. If forces are designed to respond at machine speeds, they may also escalate a conflict at machine speeds. Is there adequate time and are there deescalatory procedures to avoid unintended escalation? What is the risk of not responding to an attack where decisionmaking is happening at machine speeds? If both sides are risk-averse in this regard, are there incentives to seek a first-strike advantage?

Another set of issues arises from the difficulty of directly observing the true capabilities of a force that incorporates significant amounts of AI or that relies heavily on autonomous systems when these observations largely happen in peacetime. How will the decisionmaking change once conflict begins? What systems will stay humans-in-the-

---

[20] Angus Loten, "Growth of IoT Pushing Tech Innovation, Spending," *Wall Street Journal*, October 20, 2017.

[21] Isaac R. Porche III, Bradley Wilson, Erin-Elizabeth Johnson, Shane Tierney, and Evan Saltzman, *Data Flood: Helping the Navy Address the Rising Tide of Sensor Information*, Santa Monica, Calif.: RAND Corporation, RR-315-NAVY, 2014, p. 5.

loop? Will some processes become more autonomous, with significantly faster OODA loops? Will forces that had slower, centralized decisionmaking during peacetime have faster, decentralized reactions during conflict? These are some of the many unanswered questions that two sides can only guess at if they have not yet gone to war with each other.

Considerations When Building Autonomous Forces

There are also considerations to keep in mind when building autonomous systems into force structure. Heterogeneity in a nation's systems is inescapable. There are also trade-offs to designing forces that can affect deterrence. A world where countries are investing significantly in AI and autonomous systems may also make it harder to judge credible threats. It could also lead to arms-race instability.

Heterogeneity Is Inescapable

One inescapable feature of force development is that a nation's military forces, unless reshaped or transformed by some catastrophic or cataclysmic event, are typically incrementally evolving from some existing force architecture and force mix to some future architecture and mix. This means that, at any point in time, the forces will include a mix of older, legacy forces and newer, modernized ones. Thus, even after advanced AI and autonomous systems have achieved a critical mass within the force, the force architecture will still be heterogenous—a potential mixture of manned-unmanned, humans-in-the-loop, humans-on-the-loop, and humans-out-of-the-loop procedures, guidelines, practices, and doctrine.

The same country may have very different mixes in different domains or mission areas, achieving overmatch against an adversary in one area but lagging behind that same adversary in others. This has the potential to make the calculus on first-strike advantage (or disadvantage) difficult or impossible to assess with any confidence. The new-versus-legacy mix for any military patchwork of systems and procedures may also make the entire system of human control difficult for planners and decisionmakers to completely understand. It also opens up the possibility of unforeseen interactions and complications in executing intended human control or monitoring of machines.

There Are Trade-Offs in Designing Forces

Successful deterrence requires the ability to demonstrate both resolve and restraint. There are considerations in designing, developing, and fielding more-autonomous forces that should be kept in mind. Some of these decisions or applications represent trade-offs. For example:

- AI-enabled research, development, and acquisition processes may significantly shorten the development of new weapon systems that enable operational and strategic surprise. However, this application could limit an adversary's ability to make accurate risk assessments of military capabilities and possibly result in miscalculation. The trade-off here is the ability to develop and field new capabilities more quickly versus the destabilizing effects that such a sped-up process might have on strategic stability.
- Forward-deploying unmanned systems could reduce cost, footprint, and risk and shorten decision cycles. However, this might undercut the credibility of "trip wire" forces that signal commitments to allies and adversaries and reduce credibility in extended deterrence.
- Autonomous systems with the ability to apply rules of engagement to changing conditions could shorten response times to changing threat conditions and operational opportunities. However, such capabilities could come at the expense of predictable control over horizontal and vertical escalation.

Numerous other trade-offs exist and may be discovered in the development of AI and autonomous systems. With each decision, force designers should examine the ways that new capabilities would reinforce or challenge deterrence, and how new capabilities might interact with existing or anticipated future capabilities.

Judging Credible Threats May Be Difficult

Effective deterrence involves credible threats and credible capabilities. Where countries believe that AI is a force multiplier but where an adversary's true AI is not transparent or easily verifiable, it might be difficult to objectively understand an adversary's capabilities and accurately judge the credibility of a deterrent threat. This very uncertainty could increase the deterrent values of these systems if adversaries overestimate their capabilities, or decrease these systems' deterrent value if adversaries underestimate them.

AI Investment Could Lead to Arms Racing

Additionally, a continuous program on both sides of modernization, increasingly incorporating AI into force structure, and fielding increasingly autonomous systems could result in arms-race instabilities such as those experienced several times in the previous century.

CHAPTER NINE

Conclusion and Areas for Further Research

Conclusion

AI and autonomous systems have the potential to dramatically affect deterrence and escalation. The speed of machine decisionmaking, its differences from human understanding, the willingness of many countries to use autonomous systems, our relative inexperience with them, and continued developments in these capabilities are among the reasons. Current planning and development efforts have not kept pace on how to handle the potentially destabilizing or escalatory issues associated with these new technologies.

Although this report reflects an exploratory effort at examining these issues, we offer several areas for future research. We can and should avoid a future where the first time that planners think through inadvertent escalation or altered deterrence dynamics is after fielded systems are engaged in conflict.

Areas for Further Research

Conduct further work on deterrence theory and other theoretical frameworks that more explicitly consider the potential effects of AI and autonomous systems. Additional thinking is necessary to successfully manage intended deterrent activity and to avoid unintended escalatory activity. It will not be possible to stop the diffusion of AI and autonomous systems into military forces, so their implications should be accounted for in defense policy.[1] Significant work exists on nuclear deterrence, and we recommend extending that research into these new areas.

Evaluate the escalatory potential of new systems. We recommend system-level review of proposed AI implementations to assess the escalation implications that should be considered in their design, development, testing, or use. This could cover everything from the data used to train the algorithms, to recommended checks on the system after testing to monitor behavior in the field, to recommended ways to incorporate a system into the processes of humans and other machines to manage the possibility of inadvertent escalation.

Evaluate the escalatory potential of new operating concepts. In addition to understanding what may happen at the system level, better understanding is needed for how AI-enabled decision cycles and processes themselves or operating concepts using AI and autonomy might contribute to miscalculation or inadvertent escalation. Does the concept in question allow for deescalation? How? What are the mechanics of deescalation? Are certain operating concepts inherently more escalatory even as they offer operational advantages? Can those dynamics be managed satisfactorily?

Wargame additional scenarios at the operational and strategic levels. Wargaming can be a particularly useful tool to better understand unintended consequences. In this report, we explored a single scenario in one geographic area using an operational-level wargame that covered a week's worth of events. Additional wargames across a variety of scenarios, with different adversaries and allies, may offer additional insights. Strategic-level wargaming over greater time horizons may yield insights on topics such as horizontal escalation and arms-race instability.[2]

---

[1] Edward Moore Geist, "It's Already Too Late to Stop the AI Arms Race—We Must Manage It Instead," *Bulletin of the Atomic Scientists*, Vol. 72, No. 5, 2016, pp. 318–321.

[2] In the early 1980s, RAND developed a wargaming methodology that included rudimentary expert systems operation using rule-based systems and flexible scripts. The tools were used to explore possible escalation futures with varying assumptions of Russian aggressiveness. For descriptions of the approach and AI aspects, see Paul K. Davis and James A. Winnefeld, *The RAND Strategy Assessment Center: An Overview and Interim Conclusions*

Further study adversary and allied autonomous systems and philosophies of use. Deterrence and assurance are not just about one's own capabilities and perceptions, but about adversary and allied capabilities and perceptions as well. A good working understanding of both adversary and allied AI and autonomous systems and their expected employment will be paramount. Such information will hopefully improve interoperability with allies and enable U.S. planners to better anticipate adversary decision cycles.

Consider greater transparency with adversaries and allies. As during the Cold War, the world may be entering a period in which the United States, allies, partners, and adversaries need to develop new signals and mechanisms to manage escalation given the use of these new technologies. Greater transparency about how militaries are building and using such systems might support better communication and reduce misunderstanding and miscalculation.

About Utility and Development Options, Santa Monica, Calif.: RAND Corporation, RR-2945-DNA, 1982; and Randall Steeb and James Gillogly, *Design for an Advanced Red Agent for the RAND Strategy Assessment Center*, Santa Monica, Calif.: RAND Corporation, R-2977-DNA, 1982. Use of AI in tactical and strategic wargaming simulation has subsequently expanded, and is now well established with such systems as One Semi-Automated Forces (OneSAF), Joint Conflict and Tactical Simulation (JCATS), and Combat XXI, although these systems have only begun to represent an opponent with significant autonomous systems capabilities.

General Morphological Analysis

This appendix contains an overview of the method we used to generate the future world in Chapter Four. General morphological analysis (GMA) is a problem-structuring method for modeling nonquantifiable, complex problems that was developed by astrophysicist Fritz Zwicky in the 1940s for classifying astrophysical objects. The method was used in operations research by the 1970s as a part of the problem-formulation process.[1]

Identifying Parameters

We used GMA to identify a potential set of important factors (or "parameters") that shape the complex problem in question. As discussed in Chapter Four, the key factors that we identified as potentially impacting deterrence were

- structure of the international system
- understanding the adversary
- the AI market environment
- societal experience with AI
- AI levels of sophistication
- philosophy of employment
- force structure
- levels of autonomy.

---

[1] Tom Ritchey, *Wicked Problems—Social Messes: Decision Support Modelling with Morphological Analysis*, Berlin: Springer-Verlag, 2011, pp. 8–11.

We then identified the plausible future values of these key factors (or "parameter values"). For example, for the "AI market environment" factor, we identified the following possible future values:

- The commercial sector is dominant.
- There is competition between the private and public sectors.
- The commercial sector and government take divergent paths.
- The public and private sectors work in close collaboration.
- Governments dominate AI research.

Morphological Field

Table A.1 shows the morphological field that we created during our workshop. The full morphological field is constructed by taking every factor and laying out all plausible future values that were identified.

Taking one value from each row gives one combination of factors and describes one potential future. To calculate the total number of potential futures in the solution space, we multiply the number of values for each factor. Table A.1 thus shows a morphological field with $5 \times 4 \times 5 \times 4 \times 5 \times 3 \times 4 \times 4 = 96{,}000$ potential futures. However, there are actually fewer than 96,000 futures to consider after doing a cross-consistency analysis and removing the combinations that are logically inconsistent with one another, as discussed below.

Cross-Consistency Analysis

The next step in GMA after creating a morphological field is to eliminate inconsistent combinations of values.[2] For example, a future in which there is a collapse of the state system would be inconsistent with one that where the government is dominant in the AI sector. Table A.2 shows the sections of the cross-consistency matrix for which the study team identified inconsistent combinations of factors. Red indicates

---

[2] Ritchey, 2011, pp. 49–51.

Table A.1
Morphological Field

Factor	Potential Future Value				
Structure of the international system	U.S.-centric unipolar	China-centric unipolar	Multipolar, state-centric	Bipolar system	Collapse of the state system
Understanding of the adversary	Poor understanding and nascent collection	Imperfect understanding and collection	Historically good but declining understanding and collection	Good understanding and collection	
AI market environment	Commercial sector is dominant	Competition between commercial and government	Divergent paths between commercial and government	Cooperation/ collaboration between commercial and government	Government is dominant
Societal experience with AI	Limited experience and low trust	Limited experience and high trust	High experience and low trust	High experience and high trust	
AI sophistication	Statistically programmed	Simple adaptation	Advanced learning	True deep learning	Super-intelligence
Philosophy of use	Humans in the loop	Humans on the loop	Humans out of the loop		
Force structure	All operational units receive AI	AI replaces large portions of the force	AI community within the services	Outsourced AI capability	
Level of autonomy	Rule-based autonomy	Limited autonomy in few/simple environments	Greater autonomy in more/ complex environments	Autonomous throughout all environments	

an inconsistent combination. (Black simply notes that a factor is not assessed against itself.)

Overall, we determined that higher levels of autonomy were inconsistent with lower levels of AI sophistication, low levels of trust in

Table A.2
Cross-Consistency Analysis

	Structure of the International System					Social Experience with AI				AI Sophistication					Philosophy of Employment		
	Collapse of the state system	Bipolar system	Multipolar, state-centric	Unipolar	U.S.-centric, unipolar	High experience and high trust	High experience and low trust	Limited experience and high trust	Limited experience and low trust	Super-intelligence	True deep learning	Advanced learning	Simple adaptation	Statically programmed	Singularity	Humans on the loop	Humans in the loop
AI Market Environment																	
Commercial sector is dominant																	
Competition between commercial and government																	
Divergent paths between commercial and government																	
Cooperation/collaboration between commercial and government																	
Government is dominant	■																
Philosophy of Employment																	
Humans in the loop															■	■	■
Humans on the loop							■		■						■	■	■
Singularity														■	■	■
Force Structure																	
All operational units receive AI/autonomous systems																	
AI/autonomous systems replace large portions of the force																	
AI community within the services										■	■						
Outsourced AI capability																	
Level of Autonomy																	
Rule-based autonomy																	
Limited autonomy in few/simple environments																	
Greater autonomy in more/complex environments							■		■					■	■		
Autonomous throughout all environments							■		■			■	■	■		■	

AI, and philosophies of employment that stressed human control and oversight. We also assumed that militaries were unlikely to outsource extremely high levels of capable AI, such as superintelligence, to outside companies and would instead keep such AI in house.

Selecting Combinations

As a team, our next step was to vote for pairs of factor values that we found particularly interesting for further exploration through a wargame. For example, one team member voted for a pairing where the level of autonomy was "autonomous throughout all environments" but understanding of the adversary was characterized by "poor understanding and nascent collection." The rationale here was that this particular combination could provide room for significant miscalculation and inadvertent escalation.

After aggregating the votes for interesting pairings and discussing what nominated pairings could be combined, we chose one particular combination of values for each factor. Table A.3 illustrates the world that we chose and discussed in Chapter Four.

This method can create a set of possible future worlds to explore within the context of a problem. There are associated methods to assist in creating diverse scenario sets that span the challenge space, rather than clustering near expected futures with a few edge cases thrown in.[3]

---

[3] Hendrick Carlsen, E. Anders Eriksson, Karl Henrik Dreborg, Bengt Johansson, and Örjan Bodin, "Systematic Exploration of Scenario Spaces," *Foresight*, Vol. 18, No. 1, 2016, pp. 59–75.

Table A.3
One Future World

Factor	Potential Future Value				
Structure of the international system	U.S.-centric unipolar	China-centric unipolar	Multipolar, state-centric	Bipolar system	Collapse of the state system
Understanding of the adversary	Poor understanding and nascent collection	Imperfect understanding and collection	Historically good but declining understanding and collection	Good understanding and collection	
AI market environment	Commercial sector is dominant	Competition between commercial and government	Divergent paths between commercial and government	Cooperation/ collaboration between commercial and government	Government is dominant
Societal experience with AI	Limited experience and low trust	Limited experience and high trust	High experience and low trust	High experience and high trust	
AI sophistication	Statistically programmed	Simple adaptation	Advanced learning	True deep learning	Super-intelligence
Philosophy of use	Humans in the loop	Humans on the loop	Humans out of the loop		
Force structure	All operational units receive AI	AI replaces large portions of the force	AI community within the services	Outsourced AI capability	
Level of autonomy	Rule-based autonomy	Limited autonomy in few/simple environments	Greater autonomy in more/complex environments	Autonomous throughout all environments	

References

Allison, Graham, and Philip Zelikow, *Essence of Decision: Explaining the Cuban Missile Crisis*, 2nd ed., New York: Longman, 1999.

AlphaGo, directed by Greg Kohs, distributed by Moxie Pictures and Reel as Dirt, 2017.

Anderson, James M., Nidhi Kalra, Karlyn D. Stanley, Paul Sorensen, Constantine Samaras, and Oluwatobi A. Oluwatola, *Autonomous Vehicle Technology: A Guide for Policymakers*, Santa Monica, Calif.: RAND Corporation, RR-433-2-RC, 2016. As of October 18, 2019:
https://www.rand.org/pubs/research_reports/RR443-2.html

Barnes, Julian E., and Josh Chin, "The New Arms Race in AI," *Wall Street Journal*, March 2, 2018.

Barsade, Itai, and Michael C. Horowitz, "Artificial Intelligence Beyond the Superpowers," *Bulletin of the Atomic Scientists*, August 16, 2018. As of October 18, 2019:
https://thebulletin.org/2018/08/the-ai-arms-race-and-the-rest-of-the-world/

Bendett, Samuel, "Russia Is Poised to Surprise the U.S. in Battlefield Robotics," *Defense One*, January 25, 2018a. As of June 20, 2018:
https://www.defenseone.com/ideas/2018/01/russia-poised-surprise-us-battlefield-robotics/145439/print/

Bendett, Samuel, "Here's How the Russian Military Is Organizing to Develop AI," *Defense One*, July 20, 2018b. As of June 14, 2019:
https://www.defenseone.com/ideas/2018/07/russian-militarys-ai-development-roadmap/149900/?oref=d1-related-article

Bendett, Samuel, "Putin Orders Up a National AI Strategy," *Defense One*, January 31, 2019. As of June 14, 2019:
https://www.defenseone.com/technology/2019/01/putin-orders-national-ai-strategy/154555/

Bostrom, Nick, *Superintelligence: Paths, Dangers, Strategies*, Oxford, UK: Oxford University Press, 2014.

Boyd, John, *A Discourse on Winning and Losing,* edited by Grant T. Hammond, Maxwell Air Force Base, Ala.: Air University Press, 2018.

Boyle, Michael J., "The Race for Drones," *Orbis,* November 24, 2014, pp. 76–94.

Brecher, Michael, and Jonathan Wilkenfeld, *A Study of Crisis,* Ann Arbor, Mich.: University of Michigan Press, 1997.

Brooks, Linton, and Mira Rapp-Hooper, "Extended Deterrence, Assurance, and Reassurance in the Pacific During the Second Nuclear Age," in Ashley J. Tellis, Abraham M. Denmark, and Travis Tanner, eds., *Strategic Asia 2013–14: Asia in the Second Nuclear Age,* Seattle, Wash.: National Bureau of Asian Research, 2013.

Bughin, Jacques, Eric Hazan, Sree Ramaswamy, Michael Chui, Tera Allas, Peter Dahlstrom, Nicolaus Henke, and Monica Trench, *Artificial Intelligence: The Next Digital Frontier?* McKinsey Global Institute discussion paper, June 2017.

Carlsen, Hendrick, E. Anders Eriksson, Karl Henrik Dreborg, Bengt Johansson, and Örjan Bodin, "Systematic Exploration of Scenario Spaces," *Foresight,* Vol. 18, No. 1, 2016, pp. 59–75.

Chen Hanghui, "Artificial Intelligence: Disruptively Changing the 'Rules of the Game,'" *China National Defense News,* March 18, 2016. As of June 20, 2018 (Chinese language): http://www.81.cn/jskj/2016-03/18/content_6966873_2.htm

Cimbala, Stephen J., *The Dead Volcano: The Background and Effects of Nuclear War Complacency,* Westport, Conn.: Praeger, 2002.

Colby, Elbridge A., and Michael S. Gerson, eds., *Strategic Stability: Contending Interpretations,* Carlisle Barracks, Pa.: U.S. Army War College Press, 2013.

Cummings, M. L., *Artificial Intelligence and the Future of Warfare,* London: Chatham House, The Royal Institute of International Affairs, January 2017.

Curry, John, and Tim Price, *Matrix Games for Modern Wargaming: Developments in Professional and Educational Wargames Innovations in Wargaming,* Vol. 2, Lulu.com, 2014.

Davis, Lynn E., Michael J. McNerney, James Chow, Thomas Hamilton, Sarah Harting, and Daniel Byman. *Armed and Dangerous? UAVs and U.S. Security,* Santa Monica, Calif.: RAND Corporation, RR-449-RC, 2014. As of October 18, 2019: https://www.rand.org/pubs/research_reports/RR449.html

Davis, Paul K., and James A. Winnefeld, *The RAND Strategy Assessment Center: An Overview and Interim Conclusions About Utility and Development Options,* Santa Monica, Calif.: RAND Corporation, RR-2945-DNA, 1982. As of October 18, 2019: https://www.rand.org/pubs/reports/R2945.html

De Spiegeleire, Stephan, Matthijs Maas, and Tim Sweijs, *Artificial Intelligence and the Future of Defense: Strategic Implications for Small- and Medium-Sized Force Providers*, The Hague: Center for Strategic Studies, 2017.

Defense Science Board, *The Role of Autonomy in DoD Systems*, Washington, D.C.: Office of the Under Secretary of Defense for Acquisition, Technology, and Logistics, July 2012.

Defense Science Board, *Report of the Defense Science Board Summer Study on Autonomy*, Washington, D.C.: Office of the Under Secretary of Defense for Acquisition, Technology and Logistics, June 2016.

Department of Defense Directive 3000.09, *Autonomy in Weapons Systems*, Washington, D.C.: U.S. Department of Defense, November 21, 2012.

Desai, Munjal, *Modeling Trust to Improve Human-Robot Interaction*, dissertation, University of Massachusetts Lowell, 2012.

Fisher, Max, "The Forgotten Story of Iran Air Flight 655," *Washington Post*, October 16, 2013.

Fogarty, William M., *Formal Investigation into the Circumstances Surrounding the Downing of Iran Air Flight 655 on July 3, 1988*, DTIC AD-A203 577, July 28, 1988.

Freedberg, Sydney J., Jr., "Centaur Army: Bob Work, Robotics, and the Third Offset Strategy," *Breaking Defense*, November 9, 2015. As of June 20, 2018:
https://breakingdefense.com/2015/11/
centaur-army-bob-work-robotics-the-third-offset-strategy/

Freedberg, Sydney J., Jr., "Why a 'Human in the Loop' Can't Control AI: Richard Danzig," *Breaking Defense*, June 1, 2018a. As of July 2, 2018:
https://breakingdefense.com/2018/06/
why-a-human-in-the-loop-cant-control-ai-richard-danzig/

Freedberg, Sydney J., Jr., "Joint Artificial Intelligence Center Created Under DoD CIO," *Breaking Defense*, June 29, 2018b. As of June 14, 2019:
https://breakingdefense.com/2018/06/
joint-artificial-intelligence-center-created-under-dod-cio/

Fuhrmann, Matthew, and Michael C. Horowitz, "Droning On: Explaining the Proliferation of Unmanned Aerial Vehicles," *International Organization*, Vol. 71, No. 2, Spring 2017, pp. 397–418.

Geist, Edward, "It's Already Too Late to Stop the AI Arms Race—We Must Manage It Instead." *Bulletin of the Atomic Scientists*, Vol. 72, No. 5, 2016, pp. 318–321.

Geist, Edward, and Andrew J. Lohn, *How Might Artificial Intelligence Affect the Risk of Nuclear War?* Santa Monica, Calif.: RAND Corporation, PE-296-RC, 2018. As of October 18, 2019:
https://www.rand.org/pubs/perspectives/PE296.html

Geneva International Centre for Humanitarian Demining and Stockholm International Peace Research Institute, *Global Mapping and Analysis of Anti-Vehicle Mine Incidents in 2017*, April 2018. As of June 18, 2018: https://www.gichd.org/fileadmin/GICHD-resources/rec-documents/Brochure_AVM_2017_web.pdf

Gerson, Michael S., "Conventional Deterrence in the Second Nuclear Age," *Parameters*, Autumn 2009, pp. 32–48.

Goldstein, Avery, "First Things First: The Pressing Danger of Crisis Instability in U.S.-China Relations," *International Security*, Vol. 37, No. 4, Spring 2013, pp. 49–89.

Grady, John, "Mine Warfare in the Civil War," Army History Center, December 9, 2016. As of June 13, 2019: https://armyhistory.org/mine-warfare-in-the-civil-war/

Gunderson, J. P., and L. F. Gunderson, "Intelligence ≠ Autonomy ≠ Capability," Gamma Two, Inc., January 2004.

Harknett, Richard J., "The Logic of Conventional Deterrence and the End of the Cold War," *Security Studies*, Vol. 4, No. 1, Autumn 1994, pp. 86–114.

Healey, Denis, *The Time of My Life*, London: Michael Joseph, 1989.

Helgason, Guðmundur, "The Torpedoes," uboat.net, 2018. As of June 27, 2018: https://uboat.net/technical/torpedoes.htm

Horowitz, Michael C., Gregory C. Allen, Edoardo Saravalle, Anthony Cho, Kara Frederick, and Paul Scharre, *Artificial Intelligence and International Security*, Washington, D.C.: Center for a New American Security, July 2018.

Horowitz, Michael C., Sarah E. Kreps, and Matthew Fuhrmann, "Separating Fact from Fiction in the Debate over Drone Proliferation," *International Security*, Vol. 41, No. 2, Fall 2016, pp. 7–42.

Huang, Hui-Min, Elena R. Messina, and James S. Albus, "Toward a Generic Model for Autonomy Levels for Unmanned Systems (ALFUS)," National Institute of Standards and Technology, Intelligent Systems Division, August 2013.

Ilachinski, Andrew, *Artificial Intelligence & Autonomy: Opportunities and Challenges*, Arlington, Va.: CNA, 2017.

Jee, Charlotte, and Christina Mercer, "Driverless Car News: The Great Driverless Car Race: Where Will the UK Place?" *Tech World*, November 20, 2017.

Jervis, Robert. "Cooperation Under the Security Dilemma," *World Politics*, Vol. 30, January 1978, pp. 167–214.

Kahn, Herman, *Thinking the Unthinkable*, New York: Avon Books, 1962.

Kania, Elsa B., *Battlefield Singularity: Artificial Intelligence, Military Revolution, and China's Future Military Power*, Washington, D.C.: Center for a New American Security, 2017.

Klahr, Philip, and Donald A. Waterman, *Artificial Intelligence: A RAND Perspective*, Santa Monica, Calif.: RAND Corporation, P-7172, 1986. As of May 22, 2018: https://www.rand.org/pubs/papers/P7172.html

Klein, Mike, "Google's AlphaZero Destroys Stockfish in 100-Game Match," Chess.com, December 6, 2017. As of June 20, 2018: https://www.chess.com/news/view/google-s-alphazero-destroys-stockfish-in-100-game-match

Koch, Christof, "How the Computer Beat the Go Master," *Scientific American*, March 19, 2016. As of July 2, 2018: https://www.scientificamerican.com/article/how-the-computer-beat-the-go-master/

Kurzweil, Ray, *The Singularity Is Near: When Humans Transcend Biology*, New York: Penguin Books, 2005.

"Landmines Killed More Than 2,000 People in 2016," *Deutsche Welle*, December 14, 2017. As of June 13, 2019: https://www.dw.com/en/landmines-killed-more-than-2000-people-in-2016/a-41794852

Lee, John D., and Katrina A. See, "Trust in Automation: Designing for Appropriate Reliance," *Human Factors*, Vol. 46, No. 1, March 2004, pp. 50–80.

Levine, Robert, Thomas Schelling, and William Jones, *Crisis Games 27 Years Later: Plus C'est Déjà Vu*, Santa Monica, Calif.: RAND Corporation, P-7719, 1991. As of October 18, 2019: https://www.rand.org/pubs/papers/P7719.html

Long, Austin, ed., *Deterrence—From Cold War to Long War: Lessons from Six Decades of RAND Research*, Santa Monica, Calif.: RAND Corporation, MG-636-OSD/AF, 2008. As of October 18, 2019: https://www.rand.org/pubs/monographs/MG636.html

Loten, Angus, "Growth of IoT Pushing Tech Innovation, Spending," *Wall Street Journal*, October 20, 2017.

Markman, Jon, "This Is Why You Need to Learn About Edge Computing," *Forbes*, April 3, 2018.

Mazarr, Michael J., *Understanding Deterrence*, Santa Monica, Calif.: RAND Corporation, PE-295, 2018. As of October 18, 2019: https://www.rand.org/pubs/perspectives/PE295.html

McCarthy, John, *What Is Artificial Intelligence?* Stanford, Calif.: Stanford University Computer Science Department, November 2007. As of May 22, 2018: http://www-formal.stanford.edu/jmc/whatisai.pdf

Mehta, Aaron. "AI Makes Mattis Question 'Fundamental' Beliefs About War," *C4ISRNet*, February 17, 2018. As of June 20, 2018:
https://www.c4isrnet.com/intel-geoint/2018/02/17/
ai-makes-mattis-question-fundamental-beliefs-about-war/

Milivojevic, Dejan, "US WW2 Sub Sunk Itself When Its Own Torpedo Made a Full Circle & Struck It," *War History Online*, January 30, 2019. As of June 13, 2019:
https://www.warhistoryonline.com/instant-articles/ss-tullibee-us-wwii-submarine.html

Miller, James N., Jr., and Richard Fontaine, *A New Era in U.S.-Russian Strategic Stability*, Cambridge, Mass.: Belfer Center, Harvard Kennedy School and Center for New American Security, September 2017.

Minsky, Marvin, "Steps Toward Artificial Intelligence," *Proceedings of the IRE*, Vol. 49, No. 1, 1961, pp. 8–30.

"MK 48 Mod 7 Common Broadband Advanced Sonar System (CBASS) Heavyweight Torpedo," *Naval Technology*, undated. As of June 13, 2019:
https://www.naval-technology.com/projects/mk-48-mod-7-common-broadband-advanced-sonar-system-cbass-heavyweight-torpedo/

Morgan, Forrest E., Karl P. Mueller, Evan S. Medeiros, Kevin L. Pollpeter, and Roger Cliff, *Dangerous Thresholds: Managing Escalation in the 21st Century*, Santa Monica, Calif..: RAND Corporation, MG-614-AF, 2008. As of October 18, 2019:
https://www.rand.org/pubs/monographs/MG614.html

Mozur, Paul, "Google's AlphaGo Defeats Chinese Go Master in Win for A.I.," *New York Times*, May 23, 2017.

Newell, Allen, and Herbert Alexander Simon, "GPS: A Program That Simulates Human Thought," in Edward A. Feigenbaum and Julian Feldman, eds., *Computers and Thought*, R. Oldenbourg KG, 1963, pp. 279–293.

Nilsson, Nils J., *Artificial Intelligence: A New Synthesis*, Amsterdam: Elsevier, 1998.

Osinga, Frans, *Science, Strategy and War: The Strategic Theory of John Boyd*, The Netherlands: Eburon Academic Publishers, 2005.

Osoba, Osonde, and William Welser IV, *An Intelligence in Our Image: The Risks of Bias and Errors in Artificial Intelligence*, Santa Monica, Calif.: RAND Corporation, RR-1744-RC, 2017. As of October 18, 2019:
https://www.rand.org/pubs/research_reports/RR1744.html

Perrow, Charles, *Normal Accidents: Living with High-Risk Technologies*, New York: Basic Books, 1984.

Peterman, Robert W., *Mission-Type Orders: An Employment Concept for the Future*, Maxwell Air Force Base, Ala.: Air University Press, 1990.

Poole, David, Alan Mackworth, and Randy Goebel, *Computational Intelligence: A Logical Approach*, Oxford, UK: Oxford University Press, 1998.

Porche, Isaac R. III, Bradley Wilson, Erin-Elizabeth Johnson, Shane Tierney, and Evan Saltzman, *Data Flood: Helping the Navy Address the Rising Tide of Sensor Information*, Santa Monica, Calif.: RAND Corporation, RR-315-NAVY, 2014. As of October 18, 2019:
https://www.rand.org/pubs/research_reports/RR315.html

Posen, Barry R., *Inadvertent Escalation: Conventional War and Nuclear Risks*, Ithaca, N.Y.: Cornell University Press, 1991.

Powell, Robert, "Crisis Stability in the Nuclear Age," *American Political Science Review*, Vol. 83, No. 1, March 1989, pp. 61–76.

Ranjeev, Mittu, Donald Sofge, Alan Wagner, and W.F. Lawless, *Robust Intelligence and Trust in Autonomous Systems*, Boston, Mass.: Springer, 2016.

Raytheon, "Phalanx Close-In Weapon System: Last Line of Defense for Air, Land and Sea," webpage, undated. As of June 1, 2018:
https://www.raytheon.com/capabilities/products/phalanx

Rhodes, Edward, "Conventional Deterrence," *Comparative Strategy*, Vol. 19, No. 3, 2000, pp. 221–253.

Ritchey, Tom, *Wicked Problems—Social Messes: Decision Support Modelling with Morphological Analysis*, Berlin: Springer-Verlag, 2011.

Robles, Pablo, "China Plans to Be a World Leader in Artificial Intelligence by 2030," *South China Morning Post*, October 1, 2018. As of June 14, 2019:
https://multimedia.scmp.com/news/china/article/2166148/china-2025-artificial-intelligence/index.html

Russell, Stuart J., and Peter Norvig, *Artificial Intelligence: A Modern Approach*, 3rd ed., Upper Saddle River, N.J.: Prentice Hall, 2010.

Saalman, Lora., "Fear of False Negatives: AI and China's Nuclear Posture," *Bulletin of the Atomic Scientists*, April 24, 2018.

Sagan, Scott D., *The Limits of Safety: Organizations, Accidents, and Nuclear Weapons*, Princeton, N.J.: Princeton University Press, 1993.

Scharre, Paul, *Autonomous Weapons and Operational Risk*, Washington, D.C.: Center for New American Security, 2016.

Schelling, Thomas C., *The Strategy of Conflict*, Cambridge, Mass.: Harvard University Press, 1960.

Schelling, Thomas, C., *Arms and Influence*, New Haven, Conn.: Yale University Press, 1966.

Shattuck, Lawrence G., "Transitioning to Autonomy: A Human Systems Integration Perspective," briefing, Naval Postgraduate School, undated. As of June 13, 2019: https://human-factors.arc.nasa.gov/workshop/autonomy/download/presentations/Shaddock%20.pdf

Shenon, Philip, "Japanese Down Navy Plane in an Accident; Crew Is Safe," *New York Times*, June 5, 1996.

Simonite, Tom, "For Superpowers, Artificial Intelligence Fuels New Global Arms Race," *Wired*, September 8, 2017.

Singer, P. W., *Wired for War: The Robotics Revolution and Conflict in the 21st Century*, New York: Penguin Books, 2009.

Springer, Paul J., *Outsourcing War to Machines: The Military Robotics Revolution*, Praeger Security International, 2018.

Stanford University, One Hundred Year Study on Artificial Intelligence, *Artificial Intelligence and Life in 2030: Report of the 2015 Study Panel*, September 2016. As of July 2, 2018: https://ai100.stanford.edu/sites/default/files/ai100report10032016fnl_singles.pdf

Steeb, Randall, and James Gillogly, *Design for an Advanced Red Agent for the RAND Strategy Assessment Center*, Santa Monica, Calif.: RAND Corporation, R-2977-DNA, 1982. As of October 18, 2019: https://www.rand.org/pubs/reports/R2977.html

Steeb, Randall, and Morgan Kisselberg, "Counter-IADS MUM-T Exploratory Quantitative Analysis," RAND Briefing to 2018 Army Science Board Panel on MUM-T, July 20, 2018.

Stone, John, "Conventional Deterrence and the Challenge of Credibility," *Contemporary Security Policy*, Vol. 33, No. 1, 2012, pp. 108–123.

Thompson, Brittany N., "Theory of Mind: Understanding Others in a Social World," *Psychology Today*, blog post, July 3, 2017. As of October 18, 2019: https://www.psychologytoday.com/us/blog/socioemotional-success/201707/theory-mind-understanding-others-in-social-world

Trachtenberg, Marc, "The Meaning of Mobilization in 1914," in Steven E. Miller, Sean M. Lynn-Jones, and Stephen Van Evera, eds., *Military Strategy and the Origins of the First World War*, Princeton, N.J.: Princeton University Press, 1991.

Turing, Alan M., "Computing Machinery and Intelligence," *Mind*, Vol. 59, No. 236, 1950, pp. 433–460.

Turek, Matt, "Explainable Artificial Intelligence," webpage, Defense Advanced Research Projects Agency, undated. As of June 28, 2018: https://www.darpa.mil/program/explainable-artificial-intelligence

Turing, Alan M., "Computing Machinery and Intelligence," in Robert Epstein, Gary Roberts, and Grace Beber, eds., *Parsing the Turing Test: Philosophical and Methodological Issues in the Quest for the Thinking Computer*, Boston, Mass.: Springer, 2009, pp. 23–65.

U.S. Army Training and Doctrine Command, Army Capabilities Integration Center, *The U.S. Army Robotic and Autonomous Systems Strategy*, Fort Eustis, Va., 2017.

Van Evera, Stephen, "The Cult of the Offensive and the Origins of the First World War," in Steven E. Miller, Sean M. Lynn-Jones, and Stephen Van Evera, eds., *Military Strategy and the Origins of the First World War*, Princeton, N.J.: Princeton University Press, 1985.

Vinge, Vernor, "The Coming Technological Singularity: How to Survive in the Post-Human Era," paper presented at VISION-21 Symposium sponsored by the NASA Lewis Research Center and the Ohio Aerospace Institute, March 30–31, 1993. As of July 17, 2019:
https://pdfs.semanticscholar.org/3ab2/d953596fea9131e622410c64b3d114a84f0c.pdf

Waterman, Donald A., and Mark Peterson, "Rule-Based Models of Legal Expertise," *AAAI*, Vol. 1, 1980, pp. 272–275. As of October 18, 2019:
https://www.aaai.org/Library/AAAI/1980/aaai80-077.php

Werbos, P. J., "Beyond Regression: New Tools for Prediction and Analysis in the Behavioral Sciences," Ph. D. thesis, Harvard University, Cambridge, Mass., 1974.

Werbos, Paul J., "Generalization of Backpropagation with Application to a Recurrent Gas Market Model," *Neural Networks*, Vol. 1, No. 4, 1988, pp. 339–356.

Wuthnow, Joel, and Phillip C. Saunders, *Chinese Military Reports in the Age of Xi Jinping: Drivers, Challenges, and Implications*, Washington, D.C.: Institute for National Strategic Studies, National Defense University, Chinese Strategic Perspectives 10, 2013.

About the Authors

Yuna Huh Wong is a policy researcher at the RAND Corporation. Her research interests include scenario development, futures methods, wargaming, problem-structuring methods, and applied social science. She holds a Ph.D. in policy analysis from the Pardee RAND Graduate School and has served as an operations research analyst for the U.S. Marine Corps.

John M. Yurchak is a senior information scientist at the RAND Corporation who focuses on defense-related analysis. Before joining RAND, he was a 30-year career naval officer and spent the last third of his career immersed in requirements analysis, force development, and defense planning and programming. He holds an M.S. in computer science from the Naval Postgraduate School.

Robert W. Button is an adjunct senior researcher at the RAND Corporation. His research interests include artificial intelligence and simulation. He holds a Ph.D. in mathematics from Carnegie Mellon University.

Aaron Frank is a senior information scientist at the RAND Corporation. He specializes in the development of analytic tradecraft and decision-support tools for assessing complex national security issues. Frank holds a Ph.D. in computational social science from George Mason University.

Burgess Laird is a senior international researcher at the RAND Corporation. His subject-matter areas of expertise are defense strategy and force planning, deterrence, and proliferation. He holds an M.B.A. from Pepperdine University and has held senior positions within the Office of the U.S. Permanent Representative to the United Nations, the Office of the Secretary of Defense, Los Alamos National Laboratory, and the White House Office of Science and Technology Policy.

Osonde A. Osoba is an information scientist at the RAND Corporation. He applies machine learning expertise to diverse policy areas, such as health, defense, and technology policy, and recently has focused on data privacy and fairness in artificial intelligence and algorithmic systems more generally. Osoba holds a Ph.D. in electrical engineering from the University of Southern California.

Randall Steeb is a senior engineer at the RAND Corporation. He has directed research on a variety of subjects, including distributed simulation, artificial intelligence, advanced fire support systems, future air traffic control systems, and decision support systems. Steeb holds a Ph.D. in systems engineering from the University of California, Los Angeles.

Benjamin N. Harris is an adjunct defense analyst at the RAND Corporation and a student at the Massachusetts Institute of Technology, where he is pursuing a Ph.D. in political science. His research interests include emerging technologies, artificial intelligence, nuclear policy, military strategy, wargaming, and public support for war. He has a master's degree in global affairs from Tsinghua University.

Sebastian Joon Bae is a defense analyst at the RAND Corporation. His research interests include wargaming, counterinsurgency, hybrid warfare, violent nonstate actors, emerging technologies, and the nature of future warfare. He holds an M.A. in security studies from Georgetown University and has served as a wargaming analyst in the U.S. Marine Corps.